Qingtang Su

Color Image Watermarking

Advances in Computer Science

Volume 1

Qingtang Su

Color Image Watermarking

Algorithms and Technologies

DE GRUYTER

清華大學出版社
TSINGHUA UNIVERSITY PRESS

This work is co-published by Tsinghua University Press and Walter de Gruyter GmbH.

Author
Qingtang Su
School of Information and Electrical Engineering
Ludong University
Shandong, China

ISBN 978-3-11-048757-2
e-ISBN (PDF) 978-3-11-048773-2
e-ISBN (EPUB) 978-3-11-048763-3
Set-ISBN 978-3-11-048776-3
ISSN 2509-7253
e-ISSN 2509-7261

Library of Congress Cataloging-in-Publication Data
A CIP catalog record for this book has been applied for at the Library of Congress.

Bibliographic information published by the Deutsche Nationalbibliothek
The Deutsche Nationalbibliothek lists this publication in the Deutsche Nationalbibliografie; detailed bibliographic data are available on the Internet at http://dnb.dnb.de.

Typeset by Integra Software Services Pvt. Ltd.
Printing and binding: CPI books GmbH, Leck
♾ Printed on acid-free paper
Printed in Germany

www.degruyter.com

Preface

Purpose

With the rapid development of computer technology and network technology, the digital products, for example, text, image, audio, and video, have been widely spread on the Internet due to the characteristics of easily obtainable and able to copy, which also makes the copyright protection of digital color image to become more difficult and the piracy and copyright infringement become more and more serious. Therefore, whether using the color image as host image or as watermark image, the digital color image watermarking technology has already become one of the hot research fields and has been receiving more and more attention.

My goal with this book is to provide a framework of theory and application in which to conduct research and development of color image watermarking technology. This book is not intended as a comprehensive survey of the field of color image watermarking. Rather, it represents my own point of view on the subject.

The principles of watermarking are illustrated with several example algorithms and experiments. All of these examples based on different theories are implemented for color image watermarking only. The example algorithms are very simple. In general, they are not useful for real watermarking applications. Rather, each algorithm is intended to provide a clear illustration of a specific idea, and the experiments are intended to examine the idea's effect on performance.

Content and Organization

By analyzing the features and limitations of some existing works, this book has proposed some novel blind-extraction-based watermarking methods by means of some advanced technologies of image transform and matrix decomposition, in which the watermark capacity, robustness, invisibility, and algorithm efficiency have been investigated, respectively. Moreover, a considerable number of experiments have been provided to illustrate the performances of the proposed methods.

This book includes 10 chapters: Chapter 1 introduces the research background, significance, and state of art; the basic mathematical knowledge used in color image watermarking for creating the theoretical foundation of the algorithm is explained in Chapter 2; Chapter 3 addresses the basic knowledge of color digital image to create the professional foundation of color image blind watermarking technique. The detailed information of other chapters is introduced as follows.

In Chapter 4, by incorporating the advantages of spatial domain and transform domain, a new discrete cosine transform (DCT)-based blind color image watermarking algorithm in the spatial domain is proposed. According to the generating principle and the distribution feature of direct current (DC) coefficient in DCT transformation,

DC coefficients of each 8×8 block are directly calculated in spatial domain, and each watermark information bit is repeatedly embedded for four times by means of the coefficient quantization method. In this method, the watermark can be extracted from the watermarked image without the requirement of the original watermark or the original host image, and the final binary watermark is decided by the principles of first select the combination and majority vote. It has been shown from experimental results that the proposed method not only attains the higher robustness, but also has lower computation complexity.

Aiming at the problem that color watermark image has more information bits while it is not easily embedded, Chapter 5 introduces a novel dual-color image blind watermarking algorithm by utilizing state coding and integer wavelet transform (IWT). This method not only employs the feature that the IWT does not have rounding error, but also adopts the proposed state coding method that uses the nonbinary information to represent the watermark information. When embedding watermark, the state code of data set is modified to equal the hiding watermark information. Moreover, the rules of state coding are also used to extract watermark from the watermarked image without the original watermark or original host image. The simulation results show that the algorithm can embed high-capacity color watermark information into color host image.

In order to effectively enhance the invisibility of embedded color watermark, a singular value decomposition (SVD)-based color image watermarking method is proposed in Chapter 6. When embedding watermark, the watermark bit is embedded into 4×4 block by modifying the second row of the first column and the third row of the first column entries of U component after SVD. Then the embedded block is compensated by the proposed optimization operation. The embedded watermark is directly extracted from various attacked images by using the relation between the modified entries of U component without resorting to the original data. The experimental results show that the proposed algorithm not only overcomes the drawback of false-positive detection, but also attains better invisibility.

Aiming at the problem that the color watermarking method has weak robust, a color image blind watermarking algorithm based on Schur decomposition is proposed in Chapter 7. First, the theory of the Schur decomposition and the feature of decomposed image block are further analyzed. Then, the relationship between the coefficients is not only modified to embed watermark but also used to blindly extract watermark. Experimental results show that not only the invisibility of the proposed algorithm is ensured but also the robustness is significantly enhanced.

Considering the problem that some existing color image watermarking methods need more time to work, an efficient color image blind watermarking algorithm based on QR decomposition has been presented in Chapter 8. First, the color host image is divided to 4×4 nonoverlapping pixel blocks. Then each selected pixel block is decomposed by adopting QR decomposition and the first row of fourth column entry in the matrix R is quantified for embedding the watermark information. In the extraction procedure, the watermark can be extracted from the watermarked image without the original host image and the original watermark image. The simulation results show that the algorithm not only meets the requirements of the invisibility and robustness of watermark, but also has high efficiency in computation complexity.

The design of double-color image watermarking algorithm based on blind extraction is always a challenging work, Chapter 9 analyzes the features of Hessenberg matrix and proposes a color image watermarking algorithm based on Hessenberg decomposition. By using coefficient quantization technique, embed the encrypted color image watermarking information into the biggest coefficient of Hessenberg matrix; when extracting the watermark, the original host image or the original watermark image is not necessary. The experimental results show that the proposed watermarking algorithm has good performance in the aspects of watermark invisibility, robustness, and computational complexity.

The summary and prospect of blind watermarking for color digital image are presented in Chapter 10.

Obviously, this book has three properties as follows:

1. Exploring watermarking algorithms and performance analysis
2. Proposing novel algorithms with improved performance
3. Balancing the theories and industrial applications

This book can be used and referenced by the researchers in fields such as information hiding, information security, digital forensic, and so on, and can also be taken as reference book of graduate or undergraduate students in the specialties such as computer applications, information security, electronic and communication, and so on.

Contents

Acknowledgments

First, I must thank several people who have directly helped me in writing this book. Thanks to Lijun Bai for his enthusiasm and help with this book. Thanks to the editor and other workers of Tsinghua University Press who provided valuable feedback! Huanying Wang, the teacher of Wind Power School of Yantai, participated in the writing of the book, and the students of Ludong University, Zhiren Xue, Fan Liu, Shenghui Zhao, Youlei Ding, Qitong Zhang, and Jinke Wang, helped me with the input and modification of the words , deep thanks to them too!

Second, I thank the partial support of Natural Science Foundation of Shandong Province (No. ZR2014FM005), Key Research Project of Shandong Province (No. 2015GSF116001, 2014GGB01944), Shandong Province Education Department Project (No. J14LN20), Doctor Development Foundation of Ludong University (No. LY2014034), Key Research Project of Yantai City (No. 2016ZH057), Priority Academic Program Development of Jiangsu Higher Education Institutions, Jiangsu Collaborative Innovation Center on Atmospheric Environment and Equipment Technology, and Monograph Publishing Foundation of Ludong University, and deeply appreciate the support and help from the School of Information and Electrical Engineering at Ludong University!

Special thanks to anonymous referees for their valuable comments and suggestions, which lead to substantial improvements of this book.

Finally, I thank my family for their patience and support during this book: Qingtian Su, Wenliang Huang, Lin Su, and Huanghe Su.

Qingtang SU
Finish draft at Yantai, China
July, 2016

1 Introduction

As an important strategic resource, the acquisition, processing, storage, transportation, and security capability of information becomes an important part of a comprehensive national strength, and the information security has become one of the determinants of influencing national security, social stability, and economic development. Information hiding is a new emerging information security discipline, whose technique is hiding secret information in the common carrier file which is not easy to be noticed and making the secret information not found, stolen, modified, or sabotaged by ulterior persons, which ensures the security of information transmission in the network. As a new research field of information security fields, information hiding and digital watermarking have enjoyed great developments in recent years. Starting from the analysis of the multimedia information security, this chapter introduces the basic terminology, classification, and development of the information hiding technology, then demonstrates the background, basic concept and framework, common attack methods, and evaluation standard of digital watermarking technology, which is an important branch in the field of information hiding technology, and finally outlines the current research status of color image digital watermarking.

1.1 The Introduction of Information Hiding Technology

1.1.1 The Basic Terminology of Information Hiding Technology

In recent years, the rapid development of computer network technology and multimedia processing technology makes people around the world communicate more convenient and quickly. The digitization of multimedia data not only provides great convenience with the obtainment of multimedia information, but also enhances the efficiency and accuracy of information expression greatly. With the rapid development and the increasing popularity of the Internet, the exchange of multimedia information has reached an unprecedented depth and width, and its form of publishing is also increasingly plentiful. Nowadays, people can release his own work, important information, and deal network trade through the Internet, but the subsequent problems are also obvious, for example, work infringement becomes more easily and tampering with works also becomes more convenient. Therefore, how to make full use of the convenience of the Internet and protect intellectual property rights effectively has been highly valued among people. In this context, an emerging interdisciplinary discipline, information hiding was introduced. Today, information hiding, as a major means of covert communication and protection of intellectual property rights, has widely gained research and application.

Sometimes information hiding is referred to as data hiding, and its basic process is shown in Figure 1.1. Generally, people want to take the secretly hidden object as

DOI 10.1515/9783110487732-001

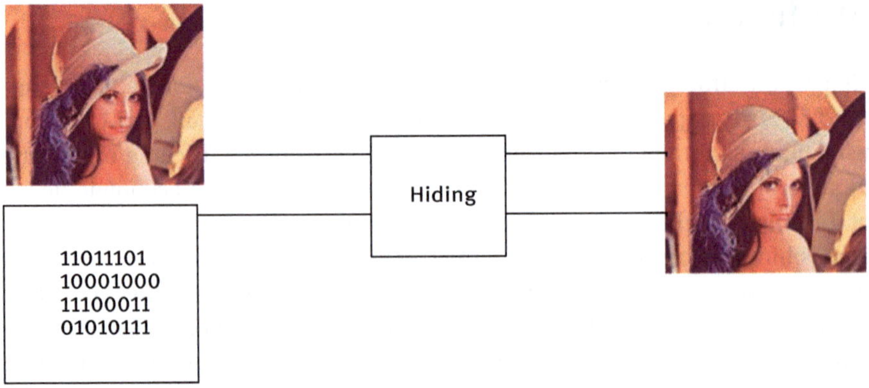

Figure 1.1: The basic process of information hiding.

embedded object; it contains specific confidential information or important information. The unclassified carrier is called carrier object which is a hidden embedded object. The meaning of "object" is widespread; it can be text, image, audio, video, software, database, and so on. The output object of information embedded process, which is called hidden object or camouflage object, has been in possession of embedded object, because it has no differences between carrier object in visual perception and auditory perception. The process of adding embedded object to the carrier object to get hidden objects is called information embedding, and the algorithm used in the process of embedding is called embedding algorithm. The inverse process of embedding information, is regaining the embedded object from the hidden object, which is called information extraction or information retrieval. The algorithm that is used in the process of extraction is called extraction algorithm. The organization or individual, who performs the process of embedding and extraction, is called the embedding and extractor, respectively.

In the information hiding system, in order to enhance security, people usually use some extra secret information to control the process of embedding and extraction, and only its holder can have access to operation, the secret information is called the hidden key. The hidden key in the process of embedding is called embedding key, and the hidden key in the process of extraction is called extraction key. Usually, the embedding key and the extraction key are identical, and the corresponding information hiding technique is called symmetrical information hiding technique, or it is called nonsymmetrical information hiding technique.

We can divide the research on information hiding into information steganography technique and hiding analyze technique. The former mainly studies the technique of secretly embedding the embedded object into the host object, while the latter mainly focuses on how to crack the information from hidden object, or how to damage the embedded information through the process of hidden object or technique that prevents the information testing. Similarly, we can call the ways or researchers to realize hidden technique hiddener, while we call the attacks on hidden system or the researchers on hidden analyze technique hidden analyzer or disguise analyzer. In addition, those above terminologies may be different in the different branches of information hidden discipline.

1.1.2 The Classification of Information Hiding Technology

1.1.2.1 The Covert Channel

Covert channel can be defined as information leakage path which is not controlled by the security policy and the scope of security policy in the system. According to the way of information transmission, covert channel can be divided into covert storage channel and covert timing channel. If a storage unit can be directly or indirectly written to a process, the unit can be directly or indirectly read by another one, then we call the channel as covert storage channel. If a process influences the real response time observed with another process by regulating the use of system resources and implements the information transmission from a process to another, then it is called as the covert timing channel.

1.1.2.2 Steganography

Steganography is an important subfield of information hiding. Unlike the protection for information contents of cryptography, steganography focuses on the hidden information itself. It comes from the Greek roots; literally, it means "steganography," which is often interpreted as that the information is hidden in other information, that is, it does not let anyone outside the program recipient to know the transfer events of information (and not just the content of the information). For example, some specific letters were written in a newspaper by using invisible ink to achieve the purpose of sending a message to spy. Modern steganography mainly refers to use the ubiquitous redundancy of computer to embed the secret data in the field of digital information processing and the computer.

1.1.2.3 The Anonymous Technology

Anonymous technology refers to a kind of technology that cannot reveal the identity and personal characteristics, and the technology is mainly applied in the network environment. Anonymity online can be divided into the sender and receiver anonymity to protect the identity of both communicating parties, respectively, and the main technologies used are anonymous remailers and network proxy.

1.1.2.4 Copyright Tag

Copyright tag is embedding the identify copyright tag information into the digital works, and the technology is an effective technique for the copyright protection of digital work. According to the difference between the content and technology which is used, the technology of copyright tag can be divided into the digital watermarking technology and the digital fingerprinting technology. Similar with money watermarking, digital watermarking technology is a special mark that embeds watermark into digital image, audio, video, and other digital products by the method of digital embedded, as the evidence of the creator's ownership of the works and evidence of identification and prosecuting the illegal infringement. At the same time, it will become the effective means of intellectual property protection and digital multimedia anticounterfeit through the watermark detection and analysis to ensure the complete reliability of digital information. Digital fingerprint technology is that producers can take different user ID or serial number as the legal copy of different embedded fingerprint works to prevent unauthorized copying and distribution. Once unauthorized copy was found, fingerprints can be recovered from this copy to determine its source [1].

1.1.3 The Development of Information Hiding Technology

Humans have had the idea for protecting information since the emergence of human culture. Cryptography and steganography are formally appeared in the mid-seventeenth century, and all derive from Greek. The earliest literature of describing the earliest information hiding is "history" written by Herodotus, known as the father of the historian, in 400 BC. An example, using waxed wooden plaque is introduced in this book: Persian Demeratus wanted to tell his Greek friend Xerxes that someone wants to violate them. At that time, a wooden plaque for writing is usually using two pieces of wax book, chained together as a book. Words are written on the wax of the book, it can be melted by wax, so the book can be reused. But the method used by Demeratus removes the wax, and writes the information on the wood, then uses wax polish on the wood. So people cannot see the hidden information under wax from the appearance. This approach worked well at first, but later been penetrated [2].

"Computer network is the mother of modern cryptography, while the Internet is the mother of the modern information hiding." The rise of computer network raises the upsurge of modern cryptography research in the 1970s and cryptography has developed into a relatively mature discipline. With the rapid development of Internet in the 1990s, the gradual maturity of multimedia technology, and the rise of e-commerce, online multimedia information has increased dramatically. If there is no network, the development of information technology will never grow so quickly; however, the openness of network and resource sharing enables the problem of network information security to become increasingly prominent. Effective measures and techniques of protecting the digital copyright are required to solve this problem, which is the main driving force of digital watermarking technique research.

At present, many digital information hiding algorithms adopt spread spectrum technology. Spread spectrum communication can be regarded as a kind of communication mode that puts information hidden in pseudorandom noise. The applications of spread spectrum communication in military have become applied more than half a century; in recent years it is widely used in civilian communication. It uses pseudorandom code that the data rate is much higher than sending information, expanding it as the information data spectrum of baseband signal, and obtaining the broadband signals with extremely low-power spectral density; thus, it is difficult to distinguish from background noise. In addition, the high frequency is good for the invisibility of the embedded information, but bad for robustness, and the low frequency is advantageous to the robustness, but it will lead to unacceptable visibility. Spread spectrum can solve this contradiction by embedding the low -frequency energy signal into each frequency band.

At present, the technology of information hiding is not yet mature; it still lacks systematic theoretical basis, fair performance test, and evaluation system, and the wide application of information hiding depends on the constant exploration and practice.

1.2 Digital Watermarking Technology

1.2.1 The Background of the Digital Watermarking Technology

With the rapid development of computer multimedia technology, people can easily use digital equipment to product, process, and store media information such as text, image, audio, video, and so on. At the same time, digital network communication is rapidly developing, which makes the distribution and transmission of information achieved "digitized" and "networked." In the analog age, people take the tape as a recording device, and the quality of pirated copies is usually lower than the original copy, that is the second copy's quality is even worse. In the digital age, the digital copy process of the song or movie can completely finish without any loss in quality. Since November 1993, the Mosaic web browser of Marc Andreessen has appeared on the Internet, and the Internet became more user-friendly; soon people gladly started to download pictures, music, and video from Internet. For digital media, Internet has become the most excellent distribution system, because it is not only cheap, but also needs no warehouse to store, and can send in real time. Therefore, digital media can be easily copied, processed, transmitted, and publicized via the use of Internet or CD-ROM. There are the security problem of digital information transmission and the copyright protection problem of digital products. How to implement effective means of copyright protection and information security in the network has caused the attention of international academic, business, and government departments. Among them, how to prevent digital products (such as electronic publications, audio, video, animation, graphics, products, etc.) from being infringement, piracy, and juggled has become one of the hot topics of the world which needs to be solved urgently.

The actual release mechanism of digital products is a lengthy process. It includes the original creators, editors, multimedia integration, pin and state officials, and so on. A simple model is given in this book, as shown in Figure 1.2. The copyright owner,

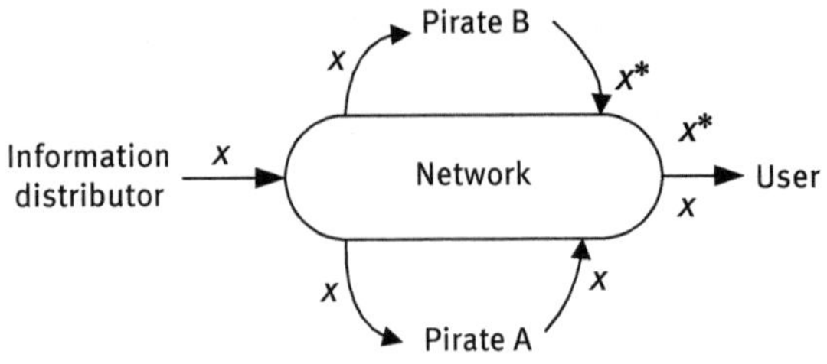

Figure 1.2: The basic model of digital production distributing in network.

editor, and repeat sales are collectively called "Information distributors" in this figure; they try to release digital product x on the network. And the "user" in the figure can also be called the consumer (customer), and he hopes to receive digital products through the network. And the "pirate" in figure is the unauthorized provider. He sends product x (such as pirate A) without permission from the legitimate copyright owners or intentionally destroyed original products (such as pirate B) and resends the untrusted version x^*; thus, user can't avoid receiving pirated copies of the x or x^* indirectly.

The illegal operation behavior, which is the pirate performs on the digital multimedia products, usually includes the following three conditions:

1. Illegal accessing, that is, illegally copying or reproducing digital products from a web site without the permission from the copyright owner.
2. Deliberately tampered, that is, pirates maliciously modify digital products or insert characteristics and resend, so that the copyright information is lost in original products.
3. Copyright damaging, that is, the pirates sell the digital products after receiving it without the permission from copyright owners.

In order to solve the problem of information security and copyright protection, first, digital product owner comes up with the encryption and digital signature technology. The encryption technology based on the private or public key can be used to control access to data; it will transform clear message into encrypted information that others don't understand. The encrypted product is still accessible, but only those who have the correct key can decrypt; in addition, you can set the password to make data cannot be read during transmission, which can provide effective protection for the data which is in the process from sending to receiving. Digital signature is used "0," "1" string to replace written signature or seal, and it has the same legal effectiveness. The digital signature can be divided into two ways: general signature and arbitration signature. The digital signature technology has been applied to test short real reliability of digital information, and has formed the digital signature standard. It gives each message a signature through the use of the private key, while the public detection algorithm is used to check the message content whether it meets the corresponding signature or not. However, this digital signature is not convenient or practical in a digital image, video, or audio application, because it needs to add a large number of signatures in the original data. In addition, with the rapid development of computer hardware and software technology, and the gradual perfection of network-based crack technology which has the capability of parallel computing, these traditional security technologies have been challenging. Alone by increasing the length of the key to enhance the reliability of the security system is no longer the only viable option, and only the personnel who is authorized to hold key information can get the encrypted information, so you can't make more people get the information they need through the public system. Also, once the information is illegally leaked, there is no direct evidence to prove that the information is illegally copied and transmitted. Furthermore, for a few people, the encryption is a challenging job, because it is difficult to prevent an encrypted file from "cropping" during the process of decryption. Therefore, we need to find more effective means that are different from traditional techniques to secure the safety of digital information and protect the copyright of digital information products.

In order to make up the deficiencies of cryptography, people begin to seek another technique to complement encryption technology, so that the decrypted content can still be protected. Digital watermarking technology may be known as a supplement technology for promising technique, because the embedded information in the digital product information will not be removed by conventional processing operations. Digital watermarking technology, on the one hand, has made up the drawbacks of cryptography, because it can provide additional protection for the decrypted files; on the other hand, digital watermarking technology has also made up the drawbacks of digital signature technology, because it can embed a lot of secret information into the original data at once. Some people have designed some kind of watermarking methods, which can remain intact in decryption, re-encryption, compression, digital-analog conversion, file format changes, and other operations. The digital watermarking technology is mainly used to prohibit the copy operation. In the copyright protection application, the watermark logo can mark the copyright owner to ensure that royalties paid together. In addition, the watermarking technology is also applied in other cases, including broadcast producer, transaction tracking, and the authenticity of identification in various fields, as well as copy control and device control.

1.2.2 The Basic Concept of Digital Watermarking

1.2.2.1 The Definition of Digital Watermarking

Watermarking is an ancient technique, and it has also been widely used in the past. A classic and well-known example is the use of invisible ink; people use invisible ink to write secret information, in order to prevent information from being perceived. Although digital watermarking technology has gotten enough attention and rapid development, but theoretically, there is no clear, uniform, and rigorous definition of a digital watermark. Cox et al. [3] defined the watermark as: "Imperceptible embedded operating behavior of information in the works." Lu [4] defined watermark as: "Digital watermarking is permanently hidden digital signals in other digital data (audio, images, video and text), and can be calculated to detect or extract for a signal confirm in the future, the digital watermark that hidden in the host image integrates with host image and does not significantly affect the visual effect of host data, so the watermarked work is still valid"; Chen et al. [5] believed that "digital watermark is digital signals or model which is permanently embedded in other data (host data) and has differential resistance, and it does not affect the availability of host data."

Generally speaking, as an effective digital copyright protection and data security maintenance technology, the digital watermarking takes full use of data redundancy, visual redundancy, or other characteristics that are ubiquitous in digital multimedia works and uses a certain embedding method to directly embed the significant tag information (digital watermark) into the digital multimedia content; as the watermarked digital multimedia work, its intrinsic value or use is not affected in any way, and the human perceptual system can't detect the embedded digital watermark information. If someone wants to extract this information, he can extract only through the designed watermark extraction algorithm or detectors to extract the embedded watermark, so that it can prove copyright ownership or certified content integrity of the digital product, then it will effectively protect the copyright of the digital multimedia and improve its security [6]. Over the past two decades, this technology has been gained widespread attention, and it has become an important research direction of information hiding.

1.2.2.2 The Features of Watermarking

1. Invisibility: Invisibility is also known as transparency and hiddenness. On the one hand, it requires that the embedded watermark must not cause significant distortion of the carrier image; on the other hand, it requires that the embedded watermark information is not invisible in subjective. Although in certain cases, copyright protection sign or logo does not need to be hidden, such as content integrity authentication, tamper detection, and recovery, but in most applications of digital watermarking, it is required the watermarked images maintain a high degree of watermark invisibility, which means that the watermarked image and the original image are almost indistinguishable in the naked eye; therefore, we should hide the watermark according to the type of information carrier and geometry and use different techniques to make it impossible for people to detect.
2. Robustness: Or hardiness, it refers to the capacities of watermarked image that can still retain watermark information after being modified intentionally or unintentionally. When the digital image is spread in the network and various transmission media, it will be interfered by adding noise, filtering, image compression, or other man-made malicious attacks. For more effective protection with copyright and other information, image watermarking algorithm should have abilities against common image processing operations [7].

3. Watermark capacity: It refers to the total information that is effectively embedded into media on the premise of satisfying the imperceptible requirements of watermark, and it is often measured by embedded binary information bits. When using a color digital image as a digital watermark, its watermark capacity is much larger than the same size binary image (at least 24 times), which undoubtedly increases the difficulty of the algorithm.
4. Security: Security refers that watermarking algorithm can guarantee the watermark information in watermarked image is secure. Not only requires that an unauthorized person can't discover digitally the watermark information of works, but also not allows extracting or destroying the information under without key situation conditions, though unauthorized person knows that the work contains a watermark and knows watermarking algorithm.

In practical application, these requirements should be taken into consideration with the designed watermarking algorithm, but there are often existing contradictions between these characteristics. Figure 1.3 shows the relationships of the robustness, invisibility, and capacity of watermark. In the process of actual application, it is impossible to meet the requirements of the best invisibility, the most robustness, and the largest capacity, but it can stress key point and make the trade-off based on actual requirements [8].

1.2.2.3 The Classification of Digital Watermarking

According to different division standards, there are many types of digital watermarking, and the specific classification is given as follows:

1. Based on the extraction (detection) method of digital watermarking
   According to the different extraction (detection) methods, the digital watermarking can be classified as non-blind watermarking algorithm [9] and blind watermarking algorithm [10–12].

   Blind watermarking algorithm requires that extracting (detecting) the watermark needs neither the participants of original data carrier, nor the help

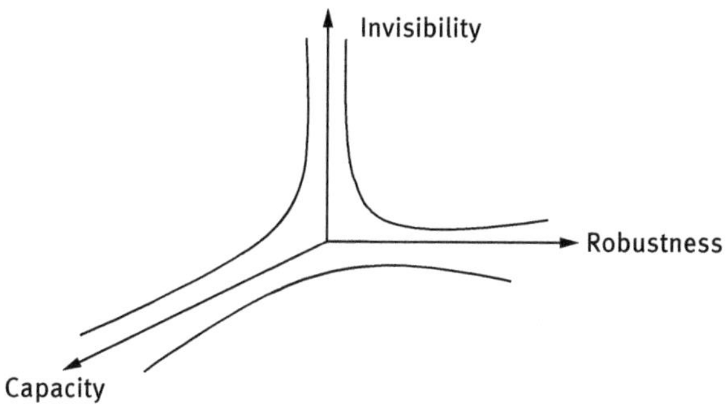

Figure 1.3: The conflict relation between the main properties of watermark.

of the original watermark; at most it needs the participation of keys. Non-blind watermarking algorithm requires the original data carrier or original watermark when extracting (detecting) watermark, so it is limited to the cost of storage and the executive efficiency. As previously mentioned, most of the current academic focuses on blind watermark in the field of academy, and it is also one of the starting points of the book.

2. Based on the capability of resisting external attacks of digital watermarking
General image processing includes compression, filtering, masking, and adding noise; in addition, there are other malicious attacks. According to the ability to resist external attack of digital watermark, digital watermarking can be divided into robust watermarking, semi-fragile watermarking, and fragile watermarking [13–16].

   For robust watermarking, it requires the ability of extracting the embedded watermark and can prove the copyright information from the watermark image whose quality has been broken out. It is mainly used for the copyright protection of media and the authenticity of identification, and so on. The fragile watermarking is different from the robust watermarking; it refers to allow watermark to be attacked by the outside world, and the damage can easily be detected. Its purpose is to analyze the extracted watermark's changes to locate and track the location and degrees of impact on the watermarked image. Semi-fragile watermark characteristic is in between them. It requires the embedded watermark can resist the general processing, but it is sensitive to malicious tampering or attack, where the fragile watermark and semi-fragile watermark, which belong to vulnerable watermarking, are mainly for integrity protection, but when the data changes, the watermark information needs to be changed accordingly, which can identify whether the original data is tampered with and how much it has been tampered with. Fragile watermarking has strong immunity when dealing with generic image processing, but it also requires strong sensitivity, which means a certain degree of distortion is allowed and the distortion can be detected [17, 18].

3. Based on the visibility of digital watermark
According to the differences in digital watermark visibility, digital watermarking can be classified as visible watermarking and invisible watermarking. The former refers to the visible digital watermark, like inserting logo on the image; it is similar to printing paper, and it is mostly used in the image to avoid these images for commercial purposes. It can also be used for video and audio. Conversely, the latter is more widely used because the human senses can't detect when embedding the invisible watermark into image, audio or video, and so on, but the invisible watermark can be found after extraction [19, 20].

4. Based on the processing mode of the host image
According to different processing ways of the host image, digital watermarking can be divided into spatial domain watermarking, transform domain watermarking, and quantization domain watermarking [21–25].

Spatial domain watermarking refers that the watermark information is embedded by directly modifying the value of the image pixel. Its characteristic is easily calculated and has a high operating efficiency, but has poor robustness relatively. Transform domain watermarking refers to embed the watermark signal into transform domain's factor. It makes good use of the human visual, auditory characteristics, and has great robustness. Discrete cosine transform (DCT), discrete wavelet transform (DWT), and discrete Fourier transform (DFT) belong to the transform domain watermarking method. Its characterization is more complicated than the spatial domain method in calculating. The quantization domain watermarking refers to select the quantization with specific structure to quantify the coefficients of carrier based on the embedded watermark information.

5. Based on the embedding objects of digital watermarking
   According to the different embedded objects, digital watermarking can be divided into image watermarking, text watermarking, audio watermarking and video watermarking when digital watermark is embedded [26–29].

1.2.2.4 The Main Application Areas of Digital Watermarking

Research on digital watermarking technology has a very important practical significance, and its main application areas are introduced as follows [30, 31]:

1. Digital product intellectual property protection
   Due to the current variety of growing online piracy, effective protection of information security and intellectual property has been attached with great importance. The digital watermarking can be a good solution to solve this problem, and it will not be eliminated when using the product through embedding the copyright information into the digital products. Smartly designed watermark can continue to exist even after compression, digital to analog conversion, change in various processing file formats, and so on.
2. Piracy tracking
   Digital watermarking can be used to track the illegal copying producing and issuing of digital products, which means that adding watermark information in each legal digital product can effectively prevent unauthorized copying and other operations on the media.
3. Image authentication
   Sometimes it needs to confirm whether the contents of digital works are tampered or not. It can be authenticated by using fragile watermarking, because fragile watermarking means the watermark cannot be detected after any small changes. If fragile watermarking can still be detected from the works in the detecting process, it can prove that the work has not been modified.
4. Counterfeit bills of business transactions
   At present, with the rapid development of image input and output devices, a variety of high-precision printers have appeared, making all kinds of counterfeit notes and counterfeit currencies to become very easy. There are many drawbacks in

anti-counterfeiting technology used before, but printing anti-counterfeiting technology which is based on digital watermarking has unique properties, watermark relies on its visual invisible hidden in the works, and its embedded content is random, consequently increasing difficulty to forgery for forgers.

1.2.3 The Basic Framework of Digital Watermarking

A complete digital watermarking scheme generally includes three parts: the watermark generation, watermark embedding, and watermark extraction or detection. Specifically, the digital watermarking technology is actually optimizing through the analysis of the carrier medium, the pretreatment of watermark, the selection of watermark embedded position, the design of embedding mode, the design extraction method of watermark, and other key aspects, in priority to meet the basic needs of the premise, seeking solutions to quasi-optimal design problem under imperceptibility, security and reliability, robustness, and other major constraints.

The basic process of digital watermark embedding is shown in Figure 1.4. Its input includes the original watermark information W, the original carrier data I, and an optional key K; its output result is the watermarked data I^*, where the watermark information can be of any form of data, such as a random sequence or a pseudorandom sequence, character or grid, binary image, gray-level or color image, three-dimensional image, and so on. Watermark generation algorithm G should be ensured uniqueness, effectiveness, irreversibility, and other attributes. Keys K can be used to strengthen security to prevent unauthorized watermark recovering or watermark extracting. The following equation defines the general process of watermark embedding:

$$I^* = E(I, W, K), \tag{1.1}$$

where I^* refers to the watermarked data, E is watermark embedding algorithm, I represents the original data, W stands for the original watermark information, and K is regarded as the set of keys. Here the key is optional, and it is generally used for watermark extraction.

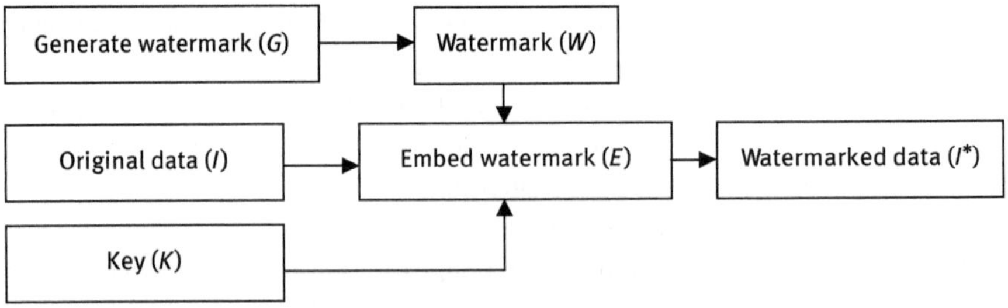

Figure 1.4: The general process of embedding watermark.

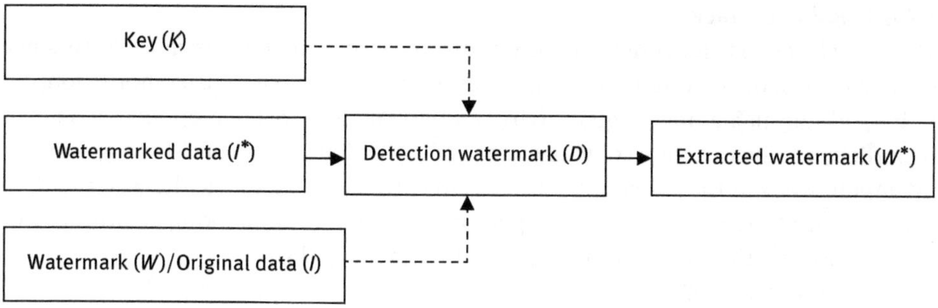

Figure 1.5: The general process of extracting watermark.

Figure 1.5 shows a general process of extracting digital watermark, and the process may need the participation of the original carrier image or original watermark. It may also don't need this information, and the watermark extraction process under different circumstances can be described as follows:

When the original carriers of data I are needed,

$$W^* = D(I^*, I, K). \tag{1.2}$$

When the original watermark W is needed,

$$W^* = D(I^*, W, K). \tag{1.3}$$

When there are no original messages,

$$W^* = D(I^*, K), \tag{1.4}$$

where W^* represents the extracted watermark, D refers to the watermark extraction algorithm, and I^* stands for a watermarked data. Equations (1.2) and (1.3) belong to non-blind extraction method, while eq. (1.4) belongs to blind manner.

1.2.4 The Attack Methods of Digital Watermarking

The attack on digital watermarking is an important measurement of digital watermarking performance. With the development of watermarking technology, the attacks on watermark are increasing. How to improve the robustness and the ability to resist attack for digital watermarking algorithm are most concerned about. A good performance watermarking algorithm should have good resistance against attack for common signal processing, geometric transformation, and malicious attacks. Therefore, in order to design a practical and effective watermarking algorithm, we must understand the various possible methods of watermarking attack.

The attacks can be divided into the following four categories:

1.2.4.1 Robust Attack

This attack means the conventional signal processing operations must experience or may experience before extracting watermarked image, such as compression, filtering, adding noise, image quantifying and enhancing, image cropping, geometric distortion, analog to digital conversion, image correction, and so on. These operations attempt to weaken the watermark strength in carrier, or to damage the watermark in carrier. Since the robustness is an important feature of watermarking algorithm, this book focuses on the common robustness attack methods [32].

1. Compression attack
   Image compression is a common attack method against the watermarked image. In general, image compression algorithm is achieving the purpose of data compression by removing redundancy in the input image signal (such as high-frequency part), and the current common image compression attacks are DCT-based JPEG compression attacks and DWT-based JPEG2000 compression attacks.
2. Adding noise attack
   The most vulnerable attack in the process of image dissemination is adding noise, so noise is also a typical attack, and it will also affect the embedded watermark. Typically, the most common noises are salt and pepper noise and Gaussian noise.
3. Median filtering attack
   Median filter is a nonlinear signal processing technology which bases on a sort of statistical theory and can effectively suppress noise. The median filtering attack of image uses a two-YCbCr window to cover the pixels in each block of image successively and to replace the medium value of the covered pixels with the pixels of the middle of the window.
4. Mosaic attack
   The so-called mosaic attack is to take the average of the pixels in a certain size template and the adjacent elements, and then assign the average value to each of pixels in the template. For example, a template of size 3×3, as shown in Figure 1.6, provides the descriptions in a simple and intuitive way.
5. Rotation, cropping, and scaling attack
   In the processing of image, a series of geometric operations need to perform on image, including image rotation, cropping, and scaling. These attacks will change the spatial position relationships between the image pixels, and bring great difficulty with the watermark extraction; so many watermarking algorithms are fragile to geometric distortion.

(a)

12	14	16
13	15	17
14	16	18

(b)

15	15	15
15	15	15
15	15	15

Figure 1.6: The example of Mosaic process: **(a)** Pixel values before processing and **(b)** Pixel values after processing.

1.2.4.2 Expression Attack

This attack is different from the robustness attack since it is not necessary to remove the embedded watermark in the digital product, which controls the content to prevent the watermark detector from detecting the presence of a watermark. For example, the expression attack fools watermark detector simply by a misaligned watermarked image; in fact, there are no changes in image pixel values in the expression attack, such as rotation, zoom, and usual affine transformation, and such attack's main idea is that watermark scheme requires watermarked image to be aligned correctly before the watermark detection.

In some of the existing image and video watermarking scheme, not only embedded expression patterns but embedded watermark in the image to resist geometric distortion, but this pattern becomes fatal weakness in the application. If the normal registration process is stopped by the attacker, then the watermark detection process will be invalidated because it can't run. For a successful expression attack, it does not need to erase or remove the watermark. In order to overcome the expression attack, watermarking software should interact with people to ensure the test will be successful. Alternatively, designing to accommodate the normal expression pattern despite achieving such intelligence in engineering is still very difficult.

1.2.4.3 Explanation Attack

Such attacks are often used fake watermark to achieve the purpose of attack. For example, the attacker does not remove the watermark, but "introduce" his own watermark in the original image, so that the watermark is meaningless, even though he does not really get the original image. In this case, there are two watermark information that can be extracted from the carrier, and attackers have the same watermark evidence of published image ownership with the owner or creator, so that the original watermark information is not representative. In the explanation attack, the pixel values of the image may be changed or not. Such attacks often require depth analysis to specific watermarking algorithm which is attacked.

1.2.4.4 Copy Attack

Copy attack is estimating watermark from a watermarked image, and then copies it to the target image to generate disguised watermarked image, and the copied watermark should be adaptive to the target image, so as to ensure its imperceptibility. Using copy to generate a fake "effective" watermark in target image needs neither watermarking algorithm nor watermark key. Copy attack is divided into three steps: the first step is to find out the estimated value of the original watermarked image; the second step is to process the estimated value; and the third step is to embed the processed watermark estimated value into target image to obtain the forged watermark image.

1.2.5 Quality Evaluation of Digital Watermarking

The quality evaluation of digital watermark includes the following two aspects: the subjective or objective evaluations of watermarked data, which is caused by the embedded watermark, and the robustness evaluations of watermark. Hence, a promising standard and more mature digital watermarking algorithm should have a good performance in at least two aspects [33].

1.2.5.1 Hiddenness

Also known as watermark invisibility, it can be understood as the ability to hide the information supplied with the digital watermark in a host image. There is a contradiction between the digital watermark information and hiddenness; with the increase of the watermark information, the quality of the image is bound to decline, and its hiddenness is also reduced accordingly. Hiddenness evaluation needs to evaluate the information account and visibility of the watermarking algorithm to demonstrate the precise relationship between the watermark information and image degradation. Hidden image evaluation can be divided into subjective evaluation and objective evaluation; both of them have their own characteristics and application situations.

1. Objectivity evaluation: It is based on the difference between the original image and the embedded watermark to evaluate the watermarked image quality. Mean square error, peak signal-to-noise ratio (PSNR) and other key indicators are commonly used as objectively appraising primary indicators of watermark carrier image distortion evaluation:

$$\text{mean square error: MSE} = \frac{1}{MN}\sum_{m,n}(I_{m,n} - I^*_{m,n})^2, \tag{1.5}$$

$$\text{signal-to-noise ratio: SNR} = \frac{\sum_{m,n} I^2_{m,n}}{\sum_{m,n}(I_{m,n} - I^*_{m,n})^2}, \tag{1.6}$$

$$\text{PSNR: PSNR} = \frac{MN\max(I^2_{m,n})}{\sum_{m,n}(I_{m,n} - I^*_{m,n})^2}, \tag{1.7}$$

where $I_{m,n}$ is the pixel of coordinates (m, n) in the original host image, $I^*_{m,n}$ refers to the pixel of coordinates (m, n) in the watermarked image, and M and N stand for the number of rows and columns of the image, respectively.

These indicators are objective image quality evaluation methods based on the all-pixel distortion statistics. However, since these two methods are all based on the pixel value comparisons of the different images, treating the same with the image of all pixels, it is only limited approximation of the human eyes' subjective senses. As natural image signal has a specific structure, and pixels have a strong correlation between them, Wang et al. research image quality issues from the standpoint of visual human system and put forward that using the structural similarity index measurement (SSIM) to measure image quality [34]. SSIM is defined as follows:

$$\mathrm{SSIM}(H, H^*) = l(H, H^*)c(H, H^*)s(H, H^*), \tag{1.8}$$

where H is the original image and H^* refers to the watermarked image:

$$\begin{cases} l(H, H^*) = \dfrac{2\mu_H\mu_{H^*} + C_1}{\mu_H^2 + \mu_{H^*}^2 + C_1}, \\ c(H, H^*) = \dfrac{2\sigma_H\sigma_{H^*} + C_2}{\sigma_H^2 + \sigma_{H^*}^2 + C_2}, \\ s(H, H^*) = \dfrac{\sigma_{HH^*} + C_3}{\sigma_H\sigma_{H^*} + C_3}. \end{cases} \tag{1.9}$$

The first item of eq. (1.9) is the comparison function of illuminance; it is used to measure the average brightness's μ_H and μ_{H^*} similarity of the two images, and the function can be obtained as the maximum value of 1 only when $\mu_H = \mu_{H^*}$. The second one is the contrast comparison function. It is used to measure the similarity of contrasts of two images, and in which σ_H and σ_{H^*} represent the standard deviation of the two images. The third one is the structure comparison function, which is used to measure the correlation coefficient of two images. σ_{HH^*} indicates the covariance between these two images; when the correlation between the two images is extremely small, the value of SSIM tends to 0, evaluating the quality of images was poor at this time. When the value of SSIM is closer to 1, the quality of images is better; when the correlation of these two images is between the above two cases, the value of SSIM is between 0 and 1. When performing the image quality evaluation experiments, in order to avoid encountering zero denominator, the numerator and denominator should add small constant terms C_1, C_2, and C_3 at the same time to correct eq. (1.9).

This book uses the SSIM of color image in eq. (1.10) to evaluate the similarity of the original color host image H and the watermarked color host image H^*:

$$\mathrm{SSIM} = \frac{\sum_{j=1}^{3} \mathrm{SSIM}_j}{3}, \tag{1.10}$$

where j represents the layer number of color host image.

2. Subjective evaluation: This method takes the person as an image viewer and makes a subjective evaluation with the pros and cons of the images. This is the approach now widely used. The observer of selecting the subjective evaluation should consider two types of people: One is not trained layman; another one is trained "experts." The so-called experts refers to an experienced observer of imaging technology; they can make judgments on the image quality to rely on their strict observation; and they can find some minor image degradation which is easy to be ignored by layman.

Table 1.1: The quality levels from 1 to 5 of ITU-R Rec.500.

Level	The damage to the quality of the image	Quality
5	Imperceptible	Excellent
4	Perceptible, not annoying	Good
3	Slightly annoying	Fair
2	Annoying	Poor
1	Very annoying	Bad

When making a subjective evaluation, the evaluation must follow a protocol that describes the complete process of testing and evaluating. This assessment is generally divided into two steps: First, dividing the distorted data set into several divisions by the order from best to worst. The second step requires the tester to grade each data set and describe the visibility according to the degraded condition. This grade may be judged based on ITU-R Rec.500 quality level in Ref. [35], which is shown in Table 1.1. The project done by European OCTALIS (Offer of Content Through Trusted Access Links) indicates that people with different experiences (such as professional photographers and researchers) hold different subjective evaluation results of the digital watermark images. The subjective evaluations have a certain practical value on the final image quality evaluation, however, in the research and development, this method is bad; hence, the actual evaluation needs to combine with objectivity evaluation method.

1.2.5.2 Robustness

Robustness refers to the ability of digital watermarking algorithm to resist various linear and nonlinear filter processing operation, and it has to have the ability to resist the usual geometric transformation and other general transformation; in a nutshell, "stability" refers to safe, and "health" stands for robust. Hence, robustness includes not only the robustness of carrying conventional treatment, but also the security to resist malicious attacks. To evaluate the robustness of watermark, it will have to be measured by a digital watermarking anti-attack capability. At present, there are many attacks watermark software, such as Stirmark, Unzign, Richard, Barnett's, Checkmark, and Optimark. Stirmark, Checkmark, and Optimark of these are more representative. In the experiments of this book, using the normalized cross-correlation (NC) shown in eq. (1.11) as the evaluation standard of robustness of binary image, it does not contain any subjective factors, so it is more impartial and reliable:

$$\text{Normalized cross-correlation: NC} = \frac{\sum_{m,n}(W_{m,n} \times W^*_{m,n})}{\sum_{m,n} W^2_{m,n}}, \tag{1.11}$$

where $W_{m,n}$ is the pixel of coordinates (m, n) in the original watermark image and $W^*_{m,n}$ refers to the pixel of coordinates (m, n) in the extracted watermark image.

For color images digital watermarking, using the following equation to calculate the NC to measure the robustness of watermarking:

$$\mathrm{NC} = \frac{\sum_{j=1}^{3}\sum_{x=1}^{m}\sum_{y=1}^{n}(W(x,y,j)\times W^*(x,y,j))}{\sqrt{\sum_{j=1}^{3}\sum_{x=1}^{m}\sum_{y=1}^{n}[W(x,y,j)]^2}\sqrt{\sum_{j=1}^{3}\sum_{x=1}^{m}\sum_{y=1}^{n}[W^*(x,y,j)]^2}}, \tag{1.12}$$

where W^* is the extracted color image watermark; W stands for the original color image watermark; $1 \le x \le m$, $1 \le y \le n$, m and n represent the row size and column size of a color image watermark, respectively, and j refers to the layers of color image watermark.

1.3 The Research Status of Color Image Digital Watermarking Technology

In the past two decades, digital watermarking technology has been deeply researched, and the watermarking algorithms that have been proposed are almost for gray-level image, but color image watermarking algorithm has not been adequately studied, because a gray-level image is easier to process than color image, and the gray-level image contains only the luminance information, and doesn't contain chrominance information, so embedding watermark in it will not create new color component [36–42]. However, color images are more common in real life as well as the dissemination of information on the Internet; it is more prone to infringement, forgery, abuse, and other bad behavior. "Tiger photo event of South China," CNN Tibet riots distorted reports, "Liu antelope," and "Zhang Flying Pigeon event" occurred in recent years are all related to the color image; therefore, color image watermarking technology gains more and more attention whether a color image is used as the host image or as a digital watermark, it becomes one of the hot spots of image watermarking technology.

At present, the study of digital watermarking technology is far more than audio, video, and digital watermarking technology, but the study of color image digital watermarking technology does not get enough attention. One of the most important reasons is that when a color image is used as a watermark, the amount of contained information is three times than the same size of the gray-level image, and it is twenty-four times than binary image, which not only increases the difficulty of color image watermarking, and the majority of existing binary image watermarking algorithm cannot be directly applied to embed the color image watermark [43]. Hence, color image watermarking technology is worthy for us to study whether a color image is used as the host image or using it as digital watermark.

In addition, the research of current color image digital watermarking is mainly non-blind watermarking [44–47], because the color images contain a large number of copyright protection information. Using non-blind watermarking technology can easily extract or embed the watermark; however, the blind watermarking technology has the advantage that non-blind watermarking technology can't match. First, non-blind watermarking needs the help of the original host image or original watermark image to detect or extract the watermark, which makes the non-blind watermarking technique to have significant limitations in practical applications. For example, the copy controlling and tracking for digital product, looking for unattacked watermarked image in the case of massive databases, and the participation of original image will make the operation more complicated and unrealistic. Second, some digital watermark detection or recovery requires a lot of data processing, which makes copious amount of raw data involved in detection or recovery to become impractical and difficult to accept, such as video watermarking applications, since the amount of data to be processed are great, and using original video is also not feasible. Finally, since the application environment of digital products is more and more networked, a small amount of data transmission and efficient detection has increasingly become a requirement which must be met by a good digital watermarking algorithm, so it can better meet the typically timeliness and security of network; therefore, the blind detection (extraction) that does not need the original data has wider application scope, and how to achieve the bind detection of color images is a hot spot problem in the field of digital watermarking [48–53].

Based on the above discussion, the research goal of this book will be blind watermarking technology which uses color images as carrier, so it not only meets the desperate requirements of copyright protection of the current popular color images in the Internet; at the same time, it will further enrich the connotation of image watermarking technology applications. If there will be a breakthrough in this technology, the content of embedded watermark can be more "colorful" and more effective for protecting the copyright performance, and it will have important applications in the field of digital media.

For what is color image digital watermarking technology, different researchers have different understandings: someone thinks that embed binary or gray-level image into color image is color image watermarking technology, others think that embed color watermark into color image is also color digital watermarking technology; therefore, we believe that both the host image and image watermark, as long as the color image-related watermarking technology all can be understood as a color image digital watermarking technology, referred to as color digital watermarking technology.

At present, the color digital watermarking researches can be divided into three categories at home and abroad: The first one is color image watermarking technology based on spatial domain; the second one is color image watermarking technology based on transform domain; and the third one is color image watermarking technology based on color quantification.

1.3.1 Research Status of Color Image Watermarking Technology Based on Spatial Domain

The earlier digital watermarking algorithms are all based on the spatial domain. Spatial domain watermark processing uses a variety of methods to directly modify the original image pixels and load the digital watermark on the original data directly. Now, several relatively typical spatial domain digital watermarking methods are demonstrated as follows.

1.3.1.1 Least Significant Bit Method

Least significant bit (LSB) algorithm is the first digital watermarking algorithm proposed by Van Schyndel et al. [54]. It is a typical spatial domain information hiding algorithm. Its idea of design is to use the limitation of human visual system (HVS) on the digital image brightness level resolution, to directly replace the LSB of digital image hide information with the hidden information, and to achieve the purpose of conveying secret information. Since the watermark signal hidden in LSB is the equivalent of adding weak signal energy, it is difficult to be detected visually and acoustically. The detection of LSB watermark is realized by the associated operation and statistical decision of test image and the watermark image. After digital image processing and image conversion, the low order of images is very easy to change. The attacker can simply filter or damage the watermark information which is added by spatial domain watermarking algorithm by deleting the image data or making mathematical transform to the digital image, so the robustness of that watermarking algorithm is very weak. Current digital watermarking software has been rarely used in the LSB algorithm [55]. However, as an information hiding approach of mass quantity data, LSB still occupies a very important position in covert communication, with the LSB idea as model [56, 57], some deformation LSB methods are proposed [58, 59], and these methods are widely used with information hiding software which is published on the Internet.

In recent years, some of the new spatial domain algorithms used machine learning techniques to embed and extract watermark. For example, neural networks, genetic algorithms, or support vector machines have been used to select the best locations or embedding watermark extraction in the spatial domain [60, 61], so the robustness of watermark is improved to resist common attacks. The deficiencies of Fu et al. [61] are that depend on the adjacent pixels having a high correlation assumption to extract the watermark, so that watermarking algorithm will be invalid when the host image is uneven.

1.3.1.2 Kutter's Methods

Kutter et al. presented the first color image watermarking algorithm [62], and the detailed instruction is given as follows: Assuming the watermark as bit information with length of X, pseudorandom sequence-based key K determines the embedded position of these information in the host image. The innovation of this method is based on the position of the watermark key formed to randomly embed watermark. The main drawback of this algorithm is that it needs to know the first two bits of watermark information when extracting watermark, and watermarked images are easy to generate virtual testing errors after geometric attacks or typical image processing. Later, several researchers such as Yu group or Tsai group suggest improving the performance of Kutter algorithm [63, 64]. The biggest difference between Yu et al. [63] and Kutter et al. [62] is the estimation of adaptive threshold in algorithm. Yu et al. [63] calculated adaptive threshold by nonlinear mapping neural network generated. However, the learning algorithm of training the neural network often converges to local optimum. In order to overcome the essential drawbacks of neural network, Tsai and Sun [64] supported to solve this problem by supporting vector machines. In the resistance of blurring or noise attack, Tsai and Sun [64] have high robustness than Yu et al. [63] and Kutter et al. [62]; on the other hand, Tsai and Sun [64] have weaker robustness in resisting geometric attacks such as rotation and scaling [36].

In recent years, much new color watermarking algorithm based on spatial domain has been proposed, and watermarking performance is improved to some extent. For example, method proposed by Huang et al. [65] directly embedded watermark into direct current (DC) component of color images in the spatial domain, and the experimental results show that except for rotation attack, the proposed algorithm has high robustness; method described in Ref. [66] divides the original host image into image blocks of different sizes, and then modifies the brightness of the block to achieve the purpose of watermark embedding. Method described in Ref. [67] also proposed a color image spatial domain algorithm based on the block, dividing the original image into nonoverlapping block with size of 8×8, and embedding watermark by modifying the intensity values of all pixels in the block. In this method, the number of watermark has to be smaller than half of the total number of block with size of 8×8. Method described in Ref. [68] raised an improved color image watermarking algorithm based on block by modifying the pixel values in each block of size 8×8 in blue component of the host image and embedding scrambled binary image information into four different positions, respectively. The experimental results show that the watermarking algorithm has good robustness against rotation, scaling, cropping, filtering, and other attacks.

1.3.2 Research Status of Color Image Watermark Technology Based on Frequency Domain

Digital watermarking technology based on transform domain tends to use similar techniques of spread spectrum image to hide the watermark information. Such techniques are generally based on the popular image transformation (based on local or global transformation), including DCT, DWT, DFT, Fourier-Mellin transform, and Hadamard transform [69].

1.3.2.1 Method of DCT Domain

DCT transform is one of the common transformations, and the first watermarking based on block DCT can be seen in methods described in Refs [70–73]. Their digital watermarking schemes are that some image blocks are randomly selected by a secret key and a triple is slightly changed to hide the binary sequence information in the middle frequency of frequency domain. The reason of choosing to encode in the middle frequency is that encoding in the high frequency can easily be destroyed by various signal processing methods, while encoding in the low frequency can be easily perceived when the low-frequency component is changed because human vision is sensitive to low-frequency component. The digital watermarking algorithm is robust for lossy compression and low-pass filter.

Cox et al. [71] proposed a digital watermarking algorithm based on image global transformation. Their important contribution is clearly that the digital watermark loaded in visual-sensitive parts of image has better robustness. Their watermarking scheme first transformed the entire image by DCT, and then loading watermark into the largest former *k* coefficients of magnitude in the DCT domain (DC component is removed), which is usually the low-frequency component. On the basis of this DCT domain embedding algorithm, Niu et al. [72] earlier proposed the image watermarking method that embeds a color digital watermark into the gray-level image. Using static image compression coding technique, color image watermark is encoded as a series of binary watermark information to realize watermark embedding. Since the watermark embedding process is based on the relationship of the DCT coefficients of the original image, watermark extracting does not need the original image. In the case of keeping the quality coefficient of 24-bit color digital image about 70%, this algorithm is compressed into an approximation of the original watermark, so it cannot really get the original watermark. Method described in Ref. [73] based on static image compression coding with DCT, and HVS, proposed a novel kind of digital watermarking algorithm that embeds a gray-level image into the original color image. Compared with the method described in Ref. [72], this algorithm can be adaptive control of embedded deep according to HVS, which is a non-blind detection algorithm. Piva et al. [74] put forward a DCT algorithm based on the statistical correlation among different color components to modify a series of coefficients of each color component for embedding watermark. Considering the sensitivity of the color components, the watermark embedding strength makes different adjustment according to the different colors.

1.3.2.2 Method of DWT Domain

DWT is a multiresolution analysis method for time/frequency signal where time/frequency domain has the ability to represent the local signal characteristics. According to the illumination masking characteristics and texture masking characteristics of the HVS, the watermark embedded in the image texture or edge is difficult to be detected. Corresponding to the wavelet transform domain of image, the image texture, edge, and other information are mainly present in some larger value of wavelet coefficients of detailed subimages, such as HH, HL, and LH. Hence, we can embed watermark information by modifying some wavelet coefficients of these detailed subimages.

Hsieh et al. [75] suggested calculating the entropy of wavelet coefficients of host image to control the invisibility and robustness of watermark. This DWT domain self-adaptive watermark has the advantages of good robustness against filtering attacks or image compression. Method described in Ref. [76] takes full advantage of HVS and uses integral lifting wavelet transform to embed the compressed color image watermark into gray-level image. Jiang and Chi [77] also proposed an algorithm that uses integer wavelet transform and HVS to embed a significant binary watermark into color image. These two algorithms effectively overcome the round-off error problem which is ubiquitous in wavelet domain watermarking algorithm. Al-Otum and Samara [78] put forward a color image blind watermarking algorithm which is robust and based on a wavelet transform. First, the algorithm forms wavelet tree of each component and uses two different components of the wavelet trees to embed watermark. The coefficient differences between the two wavelet trees are modified to ensure the embedded watermark has higher robustness, meanwhile, it should have sufficient coefficients to be selected to embed watermark, the watermark errors will be reduced to a minimum level, it improves the invisibility of the watermark. In this paper, experimental results show that the watermarked image's PSNR can reach 41.78–48.65 dB. Liu et al. [118] took full advantage of HVS of the color image and the visibility of quantization noise and raise color image watermarking technology based on blocked DWT. In order to improve the robustness and invisibility of the embedded watermark, the algorithm performs DWT on brightness component and chrominance component of the host image, selects the visual important wavelet coefficient blocks based on the threshold value of the color noise detection of color images, and decides embedding strength of the watermark information. The watermark information is embedded in the wavelet coefficients of subblocks by the quantization rule. Experimental results show that the algorithm can embed eight-color 64×64 image watermark into the 521×512 host image, and it has better watermark invisibility and robustness.

DWT can not only match the HVS better, but also it is compatible with JPEG2000 and MPEG4 compression standard. The watermarking based on wavelet transform has good visual effects and the ability to resist a variety of attacks; therefore, the digital watermark technology based on DWT is the main research direction, and it has gradually become the main tool of digital watermarking instead of DCT in transform domain.

1.3.2.3 Method of DFT Domain

DFT method is a method that uses phase of image DFT to embed information. The phase-modulated signal has stronger ability anti-interference than the amplitude-modulated signal in communication theory; similarly, using phase information to embed watermark is more robust than using amplitude information in the image, and according to the invariance of magnitude for the RST (rotation, scaling, and translation) operation, the embedded watermark can resist image RST operation, which is the effective method against the geometric attacks.

The advantage of DFT method is that the signal can be decomposed into the phase information and amplitude information, which has richer details. The DFT-based watermarking algorithms proposed by Chen [79] and Tsui et al. [80] have a strong robustness against rotation attack. In DFT domain, since the phase information has high noise immunity, it is suitable to be used to embed watermark bits. And self-adaptive phase adjustment mechanism proposed by Chen can be dynamically adjusted to the phase change and made the embedded watermark more subtle. Tsui et al. proposed color image watermarking algorithm based on quaternion Fourier transform. Because of the quaternions, the watermark is embedded as one frequency domain vector. In order to make the embedded watermark become invisible, Tsui et al. proposed watermark is embedded in the CIE L* a* b* color space. But DFT method is still relatively weak in ability to resist compression in watermarking algorithm. DFT-based watermarking algorithm is relatively small in the current watermarking algorithm.

It can be observed that the common features of transform domain watermarking algorithm are as follows: First, using the appropriate transformation method (DCT, DWT, DFT, etc.) to change information of the spatial domain of digital images into the corresponding frequency domain coefficients; second, according to the type of the hidden information, make appropriate encoding or deformation, and then establish certain rules or algorithms to modify the previously selected the coefficient sequence of frequency domain with the corresponding data of hidden information; finally, converting the frequency domain coefficient of digital image into spatial domain date by the corresponding inverse transform. Such algorithms are complicated to operate in extracting and hiding information, and can't hide a large amount of information, but it has strong anti-attack capability, which is suitable for copyright protection of digital works.

1.3.3 Research Status of Color Image Watermarking Technology Based on Color Quantization

In recent years, the watermarking technology based on color quantization is gradually being proposed, in which a quantizer with specific structure is selected to quantify the carrier coefficients according to the pre-embed watermark information. The purpose of the quantization process is to use a number of colors to represent a color image under the circumstances of minimized visual distortions [81–88].

Using the color quantization process, Pei and Chen [84] proposed to embed two watermark images into the same host image. First, the first watermark is embedded into color planes a*b* by modifying the color quantization palette, which belongs to vulnerable watermarking technology; in addition, the second watermark is embedded into the luminance component L* by modifying the quantization index of gray-level palette, which belongs to the robust watermarking technology. Tsai et al. [82] proposed a watermarking technology based on color quantization; it also performs pixel mapping and watermark embedding, and the image which has a uniform distribution of the palette has strong robustness. Recently, Chareyron et al. [89] proposed a watermarking algorithm based on color quantization of vector, and the selected pixel color quantization watermark is embedded into the xyY space of host image by modifying the color value with minimal modification. This scheme has better robustness against the geometric transformation attacks and JPEG compression attacks, but more vulnerable against the change of the color histogram.

The method, quantization index modulation (QIM), is used to quantify the color of each pixel with the same index in a host image to realize the embedding of watermark by quantifying a value in the color quantization table [88]. Chou and Wu [83] believed that the quantization process in most of QIM is not optimized according to the sensitivity of the HVS. In order to ensure the invisibility of the watermark, the color difference of the pixels of the host image and the corresponding watermark should be uniform over the entire image. Then, Chou and Wu [83] proposed that color space which has suitable quantization step size that can apply uniform quantization techniques to ensure that the color difference of adjacent elements are not detected to further improve the invisibility of watermark.

1.4 Conclusion

This chapter, starting from the analysis of the multimedia information security problems, introduces the basic terminology, classification, and development of information hiding technology, then introduces an important branch of information hiding technology field – digital watermarking technology including background knowledge, basic concepts and framework, common attack methods, and evaluation criteria, and finally introduces the research status of color image watermarking and proposes the research significance of color image digital watermarking technology.

2 Common Mathematical Knowledge Used in Digital Watermarking

The method of digital image processing is mainly divided into two categories: spatial domain analysis method and frequency domain analysis method. The former method analyzes and processes the pixels of the image directly, while the latter one transforms the image from spatial domain to frequency domain through mathematical transformation, and then analyzes and processes the image. At present, there are many mathematical transformations, such as Fourier transform, cosine transform, and wavelet transform, that can transform the image from spatial domain to frequency domain. This chapter mainly introduces some mathematical knowledge used in digital watermarking.

2.1 The Common Transforms of Image

2.1.1 The Discrete Fourier Transform

The most basic image transformation is the Fourier transform in image transformation field. By using the Fourier transform, we can solve problems in the spatial domain and frequency domain simultaneously, and Fourier transform includes continuous and discrete forms, because the images in the computer are stored through digital form, while the continuous Fourier transform is not suitable for numerical computation, so we need discrete Fourier transform (DFT) to represent the discrete information. We can also use the fast Fourier transform (FFT) to speed up the transformation.

2.1.1.1 One-Dimensional DFT

Fourier transform has a strict definition in math, and it is defined as follows [90]:

Set $f(x)$ is the function of x, if it meets the following Dirichlet's conditions:

1. a finite number of discontinuity points;
2. a finite number of pole;
3. absolutely integrable.

The Fourier transform formula is defined as

$$F(u) = \frac{1}{N}\sum_{x=0}^{N-1} f(x)\, e^{-j2\pi ux/N}, \quad u = 0, 1, 2, \ldots, N-1. \tag{2.1}$$

The inverse Fourier transform is defined as follows:

$$f(x) = \frac{1}{N}\sum_{u=0}^{N-1} F(u) e^{j2\pi ux/N}, \quad x = 0, 1, 2, \ldots, N-1, \tag{2.2}$$

where x is the time domain variable and u is the frequency domain variable.

DOI 10.1515/9783110487732-002

2.1.1.2 Two-Dimensional DFT

If the two-dimensional function $f(x, y)$ meets Dirichlet's conditions, then its two-dimensional Fourier transform formula is

$$F(u, v) = \sum_{x=0}^{M-1}\sum_{y=0}^{N-1} f(x, y)e^{-j2\pi xu/M}e^{-j2\pi yv/N}, \quad u = 0, 1, \ldots, M-1; \quad v = 0, 1, \ldots, N-1. \tag{2.3}$$

The inverse transform is defined as follows:

$$f(x, y) = \frac{1}{MN}\sum_{u=0}^{M-1}\sum_{v=0}^{N-1} F(u, v)e^{-j2\pi xu/M}e^{-j2\pi yv/N}, \quad x = 0, 1, \ldots, M-1; \ y = 0, 1, \ldots, N-1, \tag{2.4}$$

where $F(u, v)$ is called DFT coefficient of $f(x, y)$.

When $u = 0$ and $v = 0$, $F(0, 0)$ is the value of direct current (DC) component of Fourier transform (frequency is 0), when the values of v and u change from small to large, which represents the value of the alternating current (AC) component that frequency changes from low to high. If we use Matlab to solve this, the value of the matrix subscript is beginning from 1.

2.1.1.3 Three-Dimensional DFT

Three-dimensional (3D) DFT is often used in medical volume data and 3D video.

Three-dimensional continuous Fourier transform can be extended from one-dimensional and two-dimensional Fourier transform.

If $f(x, y, z)$ is continuous and integrable, and $F(u, v, w)$ is integrable as well, then there is the Fourier transform pairing with the frequency variables of u, v, w:

The forward transformation is defined as

$$F(u, v, w) = \int_{-\infty}^{+\infty}\int_{-\infty}^{+\infty}\int_{-\infty}^{+\infty} f(x, y, z)e^{-j2\pi(ux+vy+wz)}dxdydz. \tag{2.5}$$

The backward transformation is denoted as

$$f(x, y, z) = \int_{-\infty}^{+\infty}\int_{-\infty}^{+\infty}\int_{-\infty}^{+\infty} F(u, v, w)e^{-j2\pi(ux+vy+wz)}dxdydz. \tag{2.6}$$

Similar to one-dimensional and two-dimensional Fourier transform, the 3D Fourier transform formula is defined as follows:

$$F(u, v, w) = R(u, v, w) + jI(u, v, w). \tag{2.7}$$

Amplitude spectrum:

$$\left|F(u, v, w)\right| = \sqrt{R^2(u, v, w) + I^2(u, v, w)}. \tag{2.8}$$

Phase spectrum:

$$\theta(u,v,w) = \arctan\left|\frac{I(u,v,w)}{R(u,v,w)}\right|. \tag{2.9}$$

Energy spectrum:

$$P(u,v,w) = |F(u,v,w)|^2 = R^2(u,v,w) + I^2(u,v,w). \tag{2.10}$$

From the physical point of view, the amplitude spectrum represents the magnitude of the sinusoidal component; and the phase spectrum indicates the position of the sinusoidal component in the image. For the whole graph, if the phase of the sine component remains unchanged, image is basically unchanged and the amplitude spectrum has little influence on the image. In terms of the understanding of the image, the phase is more important; if we extract the image feature in the phase of the image, it will be more in line with human visual characteristics. Most of the filters do not affect the phase of the image, but only change the amplitude.

2.1.2 The Discrete Cosine Transform

Discrete cosine transform (DCT) is an important method to simplify the Fourier transform. From the properties of the Fourier transform, we can know that when the discrete real function $f(x)$, $f(x,y)$, or $f(x,y,z)$ are even functions, the calculation of transformation only has cosine terms, so the cosine transform and Fourier transform have same clear physical meaning. It can be said that the cosine transform is a special case of the Fourier transform. The DCT avoids the complex operation in the Fourier transform, which is an orthogonal transformation based on the real number. The basis vectors of matrix DCT are similar to the eigenvectors of the Toeplitz matrix that reflects the characteristics of human language and image signal, so we often think that the DCT is the optimal transform of voice and image signals. At the same time, DCT algorithm has fast speed and high precision, which is easy to be realized in the digital signal processor. Currently it occupies an important position in image processing, and it becomes core part in a series of image coding international standard (JPEG, MPEG, and H261/263).

2.1.2.1 One-Dimensional DCT

In the one-dimensional case, the forward DCT for a discrete signal $f(x)$ with length N is defined as

$$F(u) = \sqrt{\frac{2}{N}}c(u)\sum_{x=0}^{N-1} f(x)\cos\frac{\pi(2x+1)u}{2N}, \quad u = 0,1,\ldots,N-1, \tag{2.11}$$

where

$$c(u) = \begin{cases} \dfrac{1}{\sqrt{2}}, & \text{for } u = 0, \\ 1, & \text{otherwise.} \end{cases}$$

And the inverse transform is

$$f(x) = \sqrt{\frac{2}{N}} \sum_{u=0}^{N-1} c(u)F(u) \cos \frac{\pi(2x+1)u}{2N}, \quad x = 0, 1, \ldots, N-1. \tag{2.12}$$

2.1.2.2 Two-Dimensional DCT

The two-dimensional DCT can use the one-dimensional DCT to calculate, and then calculate the vertical signal since DCT is a separable function. The two-dimensional DCT of a matrix A with size of $M \times N$ is denoted as follows:

$$F(u,v) = \frac{2}{\sqrt{MN}} c(u)c(v) \left[\sum_{x=0}^{M-1} \sum_{y=0}^{N-1} f(x,y) \cos \frac{(2x+1)u\pi}{2N} \cos \frac{(2y+1)v\pi}{2M} \right],$$
$$u = 0, 1, \ldots, N-1; v = 0, 1, \ldots, M-1, \tag{2.13}$$

where

$$c(u) = \begin{cases} \sqrt{1/N}, & u = 0 \\ \sqrt{2/N}, & \le u \le N-1 \end{cases}, \quad c(v) = \begin{cases} \sqrt{1/M}, & v = 0 \\ \sqrt{2/M}, & 1 \le v \le M-1. \end{cases}$$

And the inverse transform of two-dimensional DCT can be denoted as follows:

$$f(x,y) = \frac{2}{\sqrt{MN}} c(u)c(v) \left[\sum_{u=0}^{M-1} \sum_{v=0}^{N-1} F(u,v) \cos \frac{(2x+1)u\pi}{2M} \cos \frac{(2y+1)v\pi}{2N} \right]$$
$$x = 0, 1, \ldots, M-1; \; y = 0, 1, \ldots, N-1. \tag{2.14}$$

The above-mentioned function is a separable function. When $v = 0, 1, \ldots, M-1$, then

$$f(x,y) = \sqrt{\frac{2}{M}} \sum_{v=0}^{M-1} C(v)F(u,v) \cos \left[\frac{\pi(2y+1)v}{2M} \right].$$

When $u = 0, 1, \ldots, N-1$, then

$$f(x,y) = \sqrt{\frac{2}{N}} \sum_{u=0}^{N-1} C(u)F(u,y) \cos \left[\frac{\pi(2x+1)u}{2N} \right],$$

where x, y are the sampling values in the spatial domain, and u, v are the sampling values in the frequency domain. In digital image processing, the digital image is usually represented by a square matrix, that is, $M = N$.

After the image is transformed by two-dimensional DCT, its coefficients can be divided into a DC component and a series of AC components. Among them, DC component represents the average luminance and AC components are concentrated in the main energy of the original image block. The AC components are made up of three parts, which include the low-frequency section, intermediate frequency section, and high-frequency section. The energy is concentrated in the low-frequency coefficients, the intermediate frequency coefficients gathered in a small part of the energy of the image, while the high-frequency coefficients gathered less energy. The high-frequency components of the AC components are to discard first in JPEG compression. Therefore, the algorithm embedded watermark signal in the low-frequency portion generally has good resistance against JPEG compression and anti-scaling of resampling.

2.1.2.3 Three-Dimensional DCT

The 3D DCT of 3D data A of size $M \times N \times P$ is defined as follows:

$$F(u,v,w) = c(u)c(v)c(w)\left[\sum_{x=0}^{M-1}\sum_{y=0}^{N-1}\sum_{w=0}^{P-1} f(x,y,z)\cos\frac{(2x+1)u\pi}{2M}\cos\frac{(2y+1)v\pi}{2N}\cos\frac{(2z+1)w\pi}{2P}\right],$$
$$u = 0,1,\ldots,M-1; v = 0,1,\ldots,N-1; w = 0,1,\ldots,P-1, \quad (2.15)$$

where

$$c(u) = \begin{cases}\sqrt{1/M}, & u = 0,\\ \sqrt{2/M}, & u = 1,\ldots,M-1,\end{cases}$$

$$c(v) = \begin{cases}\sqrt{1/N}, & v = 0,\\ \sqrt{2/N}, & v = 1,\ldots,N-1,\end{cases}$$

$$c(w) = \begin{cases}\sqrt{1/P}, & w = 0,\\ \sqrt{2/P}, & w = 1,\ldots,P-1.\end{cases}$$

And the inverse transform of 3D DCT is denoted as follows:

$$f(x,y,z) = \sum_{x=0}^{M-1}\sum_{y=0}^{N-1}\sum_{p=0}^{P-1}\left[c(u)c(v)c(w)F(u,v,w)\cos\frac{(2x+1)u\pi}{2M}\cos\frac{(2y+1)v\pi}{2N}\cos\frac{(2z+1)w\pi}{2P}\right], \quad (2.16)$$

where $u = 0, 1, \ldots, M-1$; $v = 0, 1, \ldots, N-1$; $w = 0, 1, \ldots, P-1$, and

$$c(u) = \begin{cases} \sqrt{1/M}, & u = 0, \\ \sqrt{2/M}, & u = 1, \ldots, M-1, \end{cases}$$

$$c(v) = \begin{cases} \sqrt{1/N}, & v = 0, \\ \sqrt{2/N}, & v = 1, \ldots, N-1, \end{cases}$$

$$c(w) = \begin{cases} \sqrt{1/P}, & w = 0, \\ \sqrt{2/P}, & w = 1, \ldots, P-1. \end{cases}$$

2.1.3 The Discrete Wavelet Transform

The procedures of digital watermark embedding and detection usually involve discrete wavelet transform (DWT), so we first introduce the continuous wavelet transform (CWT), and then focus on the two-dimensional DWT and 3D DWT [91].

The wavelet analysis method is a time-frequency localization analysis method, whose window size is fixed but its shape can be changed, and time window and frequency window can be changed. The low-frequency part is provided with high resolution in frequency and low resolution in time, so it is known as math microscope. Because of this feature, the wavelet transform has the adaptability to signal. In principle, the local traditional Fourier analysis can be used instead of wavelet analysis. Wavelet analysis is superior to Fourier transform since it has good localization properties and ability to represent local characteristic of signal in the time domain and the frequency domain, and it is very suitable in detecting the entrainment of the normal signal transient abnormal phenomena and showing its components.

Suppose $\psi(t) \in L^2(R)$, its Fourier transform $\hat{\psi}(\omega)$ is assumed to meet the admissibility condition:

$$C_\psi = \int_R \frac{|\hat{\psi}(\omega)|^2}{|\omega|} d\omega < \infty.$$

Then $\psi(t)$ is called a basic wavelet or the mother wavelet. A family of functions, as shown in eq. (2.17), can be obtained by the expansion and translation of the basic wavelet $\psi(t)$:

$$\psi_{a,b}(t) = \frac{1}{\sqrt{a}} \psi\left(\frac{t-b}{a}\right), \quad a, b \in R, a > 0. \tag{2.17}$$

The function is called continuous wavelet base function (in short named wavelet), a is the scale factor, and b is the translation factor, and they are continuous changing values.

2.1.3.1 Continuous Wavelet Transform

Definition 2.1. The CWT of any function $f(t) \in L^2(R)$ is

$$Wf(a,b) = \int_R f(t)\frac{1}{\sqrt{a}}\psi^*\left(\frac{t-b}{a}\right)dt = < f, \psi_{a,b} >. \tag{2.18}$$

If the wavelet used in the wavelet transform meets the admissibility condition, then the inverse transform exists and defined as follows:

$$f(t) = \frac{1}{C_\psi}\int_R\int_R Wf(a,b)\psi_{a,b}(t)\frac{da}{a^2}db. \tag{2.19}$$

CWT has the following important properties:

1. Linear property: a function of CWT is equal to the transformation sum of each component in the function, and the formula is expressed as follows:

   $$\text{If } f(t) = f_1(t) + f_2(t) f(t) \leftrightarrow Wf(a,b) f_1(t) \leftrightarrow Wf_1(a,b) f_2(t) \leftrightarrow Wf_2(a,b),$$

   then

   $$Wf(a,b) = Wf_1(a,b) + Wf_2(a,b).$$

2. Translation invariance: if $f(t) \leftrightarrow Wf(a,b)$, then $f(t-u) \leftrightarrow Wf(a,b-u)$.
3. Scaling covariance: if $f(t) \leftrightarrow Wf(a,b)$, then $f(ct) \leftrightarrow c^{-1/2}Wf(ca,cb)$.

The coefficients of CWT have a large amount of redundancies. In a continuous transformation of the scale a and time b, the wavelet basis function $\psi_{a,b}(t)$ has great relevance, and thus the amount of information of signal wavelet transform coefficient $Wf(a,b)$ is redundant. In most cases, we try to reduce the redundancy of the wavelet transform coefficients without losing the original signal information for improving the compression ratio. Therefore, the DWT is introduced.

2.1.3.2 The DWT

In practical applications, especially implemented on a computer, the continuous wavelet function $\psi_{a,b}(t)$ and the CWT $Wf(a,b)$ need to be discretized. Here, the so-called discretization is specific to the discrete scale factor a and the translation factor b, rather than the discrete time t.

Typically, the scale factor a and the translation factor b of discretization equations were taken as $a = 2^j$, $b_{j,k} = 2^j k$, in which $j, k \in Z$ and the corresponding discrete wavelet function $\psi_{j,k}(t)$ is defined as follows:

$$\psi_{j,k}(t) = \frac{1}{\sqrt{2^j}}\psi\left(\frac{t}{2^j} - k\right). \tag{2.20}$$

In this case, any function $f(t) \in L^2(R)$ of the DWT can be represented as

$$Wf(j,k) = \int_R f(t)\frac{1}{\sqrt{2^j}}\psi^*\left(\frac{t}{2^j} - k\right)dt =< f, \psi_{j,k} >. \tag{2.21}$$

1. Two-dimensional DWT

   Based on the one-dimensional DWT, we can easily extend wavelet transform to two-dimensional DWT. In the two-dimensional case, a two-dimensional scaling function $\phi(x,y)$ is needed, and only consideration of the case of scaling function is separable, that is

$$\phi(x,y) = \phi(x)\cdot\phi(y), \tag{2.22}$$

   where $\phi(x)$ is a one-dimensional scaling function.

   (1) Forward transform

   Starting from an image $f(z,y)$ of size $N \times N$, where N is a power of 2, the subscript j represents the resolution parameters, the scale is $2j$. If $j = 0$, then the scale of the original image is 1; it is the resolution signal in the first level 0. As j increases each time, the scale is doubled, which leaving the resolution is halved. Dimensional DWT is performed as follows: each level image is decomposed to four 1/4-size images, which are generated by the twofold interval sampling at the row and column directions after getting the inner product of the original image and the image based on wavelet.

   As shown in Figure 2.1, we can see that image wavelet transform is using the filter module to filter the image at horizontal and vertical directions; at each scale, $f_j^0(x,y)$ includes the low-frequency approximation component of the former layer, and $f_j^1(x,y)$, $f_j^2(x,y)$, and $f_j^3(x,y)$ contain the detailed components of horizontal, vertical, and high-frequency part, respectively.

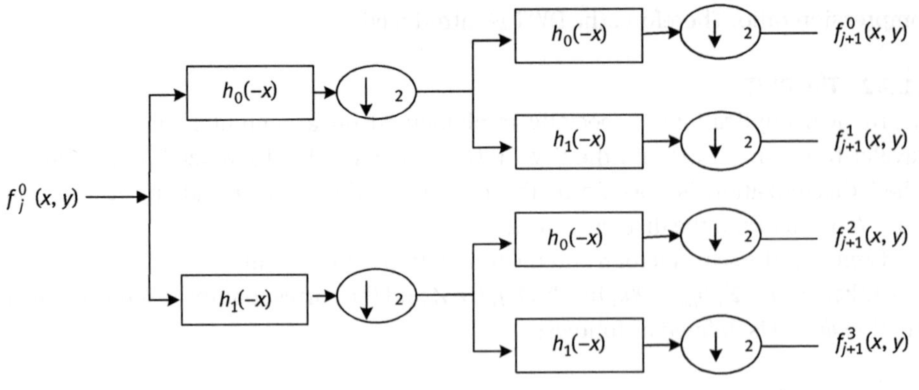

Figure 2.1: The diagram of DWT.

(2) Inverse transform
The inverse transform is similar with the forward transform process; insert a zero column in the left of each column of four arrays to increase the sampling frequency in each layer, convolving each row with the $h_0(-x)$ and $h_1(-x)$, and summing up the pair arrays of these $(N \times N)/2$ together. Then, insert a zero row on the top of each row to get the sampling frequency of the just obtained two arrays of $N \times N$, convolving each column with the $h_0(-x)$ and $h_1(-x)$. Finally, the sum of the two arrays is the rebuilt result of the layer. The implementation process is introduced in Figure 2.2.

For two-channel digital image signals, wavelet transform, which decomposes the image with multiresolution, decomposes the image into sub-images with different spaces and different frequencies. That is, the image is divided into four frequency bands, including horizontal sub-band *HL*, vertical sub-band *LH*, and diagonal sub-band *LH* (the first letter indicates the frequency in the horizontal direction and the second letter indicates the frequency of the vertical direction). By performing the multilevel resolution, sub-band *LL* can proceed by the two-dimensional DWT. Hence, Mallat decomposition is also called binary decomposition or double-frequency decomposition. The sub-band structure, as shown in Figure 2.3, is the two-level decomposed Lena image. Let N_L be the wavelet decomposition levels, then the original image is denoted as LL_0, the lowest frequency sub-band is denoted as LL_{NL} (may be abbreviated as *LL*), and other levels of sub-band are denoted as HL_K, LH_K, LH_K, HH_K, where $1 \le K \le N_L$. Mallat decomposition has many resolution features, and the image becomes $N+1$ resolution naturally after wavelet decomposition with decomposition level N_L.

Although the data amount of generated wavelet image after the wavelet transform is equal to that in the original image, the features of transformed image are different with that of the original image, which includes the image energy that mainly concentrates in the low frequency, while the horizontal, vertical, and diagonal sections contain less energy. Here, *L* represents the low-pass filter and *H* is the high-pass filter. LL_2, as two sub-band approaches of the original brightness

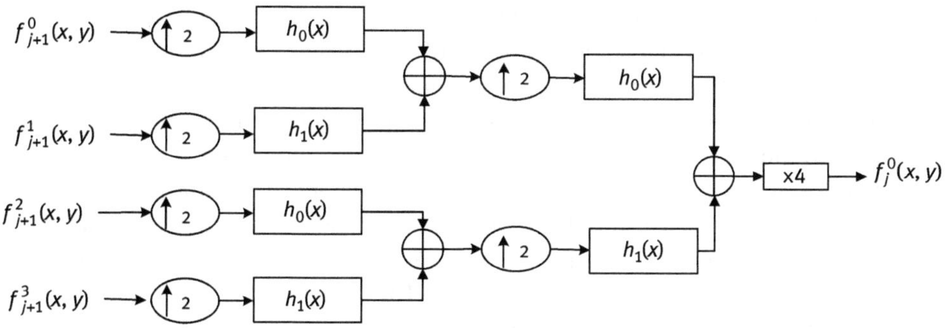

Figure 2.2: The diagram of inverse DWT.

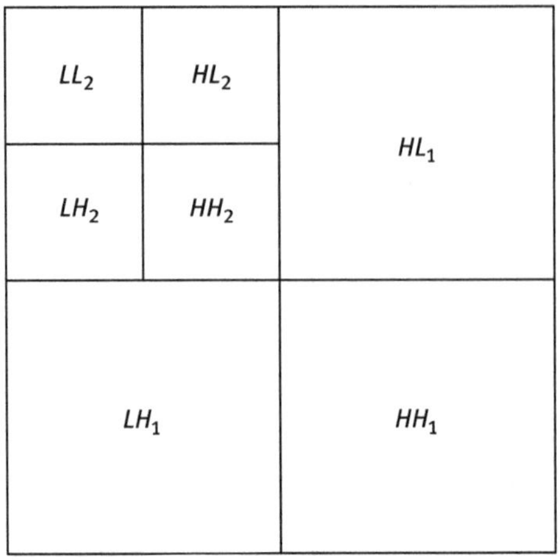

Figure 2.3: The two-level DWT of Lena image.

image, concentrated the vast majority energy of the image; the middle-frequency sub-bands HL_K, $LH_K(k \in 1, 2)$ are the details in the horizontal and vertical directions of the original image. The high-frequency sub-band $HH_K(k \in 1, 2)$ is the detailed part in the diagonal direction of the original image. The feature of multiresolution image based on wavelet transform shows that it has good spatial selectivity on direction.

According to decomposed wavelet image from Lena, sub-band LL concentrates on the vast majority energy of the original Lena image, which is approximation of the original image. Sub-band images HL, LH, and LH, respectively, maintain the details in the vertical edge of the image, the details in the horizontal edge, and the details in diagonal edge, which depict the detailed features of the image and be called the detailed sub-images. In order to improve the robustness of the watermarking, researchers often embed a watermark into a low-frequency part of the image.

2. Three-dimensional DWT

For 3D number of font data signals, wavelet transform is multiresolution decomposition for volume data by decomposing the image into sub-images X, Y, Z in different directions. Volume data is divided into eight bands by 3D wavelet transform.

One layer decomposition process of 3D wavelet is shown in Figure 2.4, where L and H represent the low-frequency components and high-frequency components, receptively, after low- and high-frequency filtering. Similar with the two-dimensional image wavelet transform, volume data is decomposed into "approximate coefficients" LLL_1 (i.e., 3D low-frequency sub-band) that represent the low-frequency feature of the volume data and "detailed coefficients" (i.e., 3D high-frequency sub-band) that represents the high-frequency feature of the volume data after 3D wavelet transform. The subscript "1" represents the first layer of the 3D DWT decomposition.

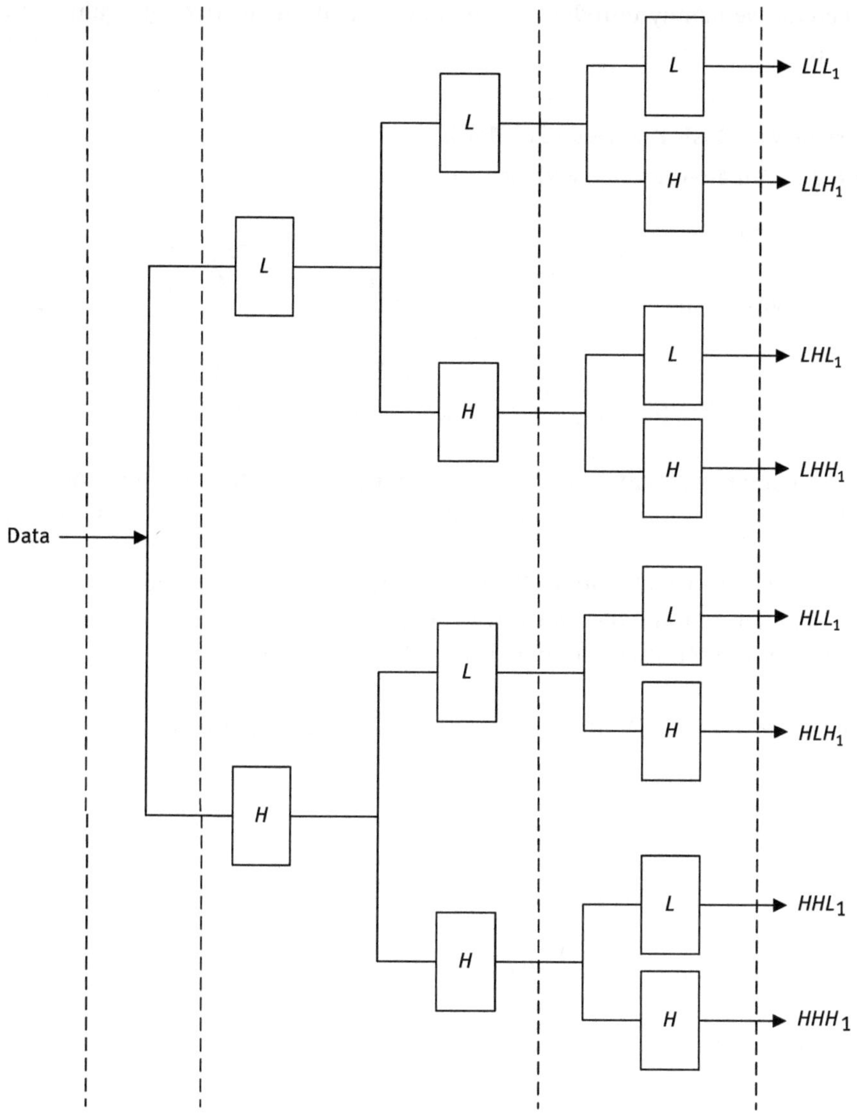

Figure 2.4: The one-level decomposition process of three-dimensional DWT.

2.2 Common Matrix Decomposition

2.2.1 Singular Value Decomposition

Digital watermarking technology generally involves matrix feature applications, such as the triangular decomposition of singular matrix applied to the embedding algorithm watermark and digital watermarking algorithm based on matrix eigenvalues.

In this section, we mainly introduce some knowledge of the matrix eigenvalue and eigenvector.

2.2.1.1 Review on Basic Knowledge of Matrix

Definition 2.2. A matrix $A = (a_{ij})$ with order n, then

$$\phi(\lambda) = \det(\lambda I - A) = \det\begin{pmatrix} \lambda - a_{11} & -a_{12} & \cdots & -a_{1n} \\ -a_{21} & \lambda - a_{22} & \cdots & -a_{2n} \\ \vdots & \vdots & \ddots & \vdots \\ -a_{n1} & -a_{n2} & \cdots & \lambda - a_{nn} \end{pmatrix} \tag{2.23}$$
$$= \lambda^n - (a_{11} + a_{22} + \cdots + a_{nn})\lambda^{n-1} + (\text{item times} \le n-2).$$

Usually, there are n roots (real root or complex root, and the complex roots are calculated by multiplicity) known as eigenvalues of A. $\lambda_{(A)}$ represents the set of all eigenvalues.

Note: When A is a real matrix, $\Psi(\lambda) = 0$ is the algebraic n equation with real coefficients, and its plural roots appear with conjugate pairs.

The following describes some conclusions about the eigenvalues:

Theorem 2.1. *Let $\lambda_i (i = 1, 2, \Lambda, n)$ is eigenvalues of matrix $A = (a_{ij})$ with order n, then:*
1. $\sum_{i=1}^{n} \lambda_i = \sum_{i=1}^{n} a_{ii} = tr(A)$ *is called locus of A.*
2. $|A| = \lambda_1 \lambda_2 \Lambda \lambda_n$.

Theorem 2.2. *Let $A \in R^{n \times n}$, then*

$$\lambda(A^T) = \lambda(A). \tag{2.24}$$

Theorem 2.3. *Suppose A is block upper triangular matrix, namely,*

$$A = \begin{bmatrix} A_{11} & A_{12} & \cdots & A_{1m} \\ & A_{22} & \cdots & A_{2m} \\ & & \ddots & \vdots \\ & & & A_{mm} \end{bmatrix},$$

where each diagonal block A_{ii} is a square matrix, then

$$\lambda(A) = \sum_{i=1}^{n} \lambda(A_{ii}). \tag{2.25}$$

Theorem 2.4. *Suppose A and B are similar matrices (namely existing a nonsingular matrix P which makes* $B = P^{-1}AP$*), then*
1. *A has the same eigenvalues with B.*
2. *If y is the eigenvector of B, then* P_y *is the eigenvector of A.*

Theorem 2.4 shows the eigenvalues of matrix A are unchanged when a similar transformation on matrix A is performed.

Definition 2.3. If the real matrix A has an eigenvalue λ whose multiplicity is k and the number of linear-independent eigenvectors of matrix A corresponding to λ is less than k, then A is called loss matrix.

A loss matrix, which doesn't have enough eigenvectors, has difficulties both in theory and calculation.

Theorem 2.5. *If* $A \in R_{n\times n}$ *is diagonalizable, then there is a nonsingular matrix P which makes*

$$P^{-1}AP = \begin{bmatrix} \lambda_1 & & & \\ & \lambda_2 & & \\ & & \ddots & \\ & & & \lambda_n \end{bmatrix}. \tag{2.26}$$

The necessary and sufficient conditions are A has n linearly independent eigenvectors.

If $A \in R_{n\times n}$ *has* $m(m \le n)$ *different eigenvalues of* $\lambda_1, \lambda_2, \Lambda, \lambda_m$*, then the corresponding eigenvectors* X_1, X_2, Λ, X_m *are linearly independent.*

Theorem 2.6 (Symmetric Matrix's Orthogonal Reduction). *Suppose* $A \in R_{n\times n}$ *is a symmetric matrix, then*
1. *The eigenvalues of A are real numbers.*
2. *A has n linearly independent eigenvectors.*
3. *There is an orthogonal matrix P which makes*

$$P^TAP = \begin{bmatrix} \lambda_1 & & & \\ & \lambda_2 & & \\ & & \ddots & \\ & & & \lambda_n \end{bmatrix}. \tag{2.27}$$

And λ_1, λ_2, Λ, λ_n *are eigenvalues of A, the column vector* u_j *which belongs to* $P = (u_1, u_2, \Lambda, u_n)$ *is unit eigenvector of A corresponding to* λ_j.

Theorem 2.7 (Gershgorin Circle Theorem).

1. *Let matrix $A = (a_{ij})$ with order n, and each of the eigenvalue of A must be in one of the following disks*

$$\left|\lambda - a_{ii}\right| \leq r_i = \sum_{\substack{j=1\\ j\neq i}}^{n} \left|a_{ij}\right| \quad (i = 1, 2, \ldots, n), \tag{2.28}$$

   or we can say all the eigenvalues of A lie in the union of n disks.
2. *If A has m disks and make up a connected union S, and S is separated from the remaining $n - m$ disks, then S contains m eigenvalues of A appropriately.*

In particular, if D_i, one of the A disks, is separated from the rest of the disks (i.e., the isolated disk), then an eigenvalue of A is precisely included in D_i.

Proof. Only (1) is given to prove. Suppose λ is an eigenvalue of A, that is $Ax = \lambda x$. Let $|x_k| = \max|x_i| = \|x\|_\infty \neq 0$, considering the kth equation of $A_x = \lambda_x$, namely,

$$\sum_{j=1}^{n} a_{kj}x_j = \lambda x_k$$

or

$$(\lambda - a_{kk})x_k = \sum_{j\neq k} a_{kj}x_j.$$

Thus,

$$\left|\lambda - a_{kk}\right| \leq \sum_{\substack{j=1\\ j\neq k}}^{n} \left|a_{kj}\right| = r_k.$$

This shows that each eigenvalue of A must lie in the disk of A, and the corresponding eigenvalue λ must lie in the kth disk (where k is the largest component of the absolute value of the corresponding feature vector).

By using the properties of similar matrix, sometimes we can get the eigenvalues of A to further evaluate, which selecting proper nonsingular matrix

$$D^{-1} = \begin{pmatrix} \alpha_1^{-1} & & & \\ & \alpha_2^{-1} & & \\ & & \ddots & \\ & & & \alpha_n^{-1} \end{pmatrix}.$$

and operating the similar transform

$$D^{-1}AD = \left(\frac{a_{ij}a_j}{a_i}\right)_{n\times n}. \tag{2.29}$$

2.2.1.2 The Power and Inverse Power Methods

Power method and inverse power method are realistic vector iteration methods to calculate eigenvalues and eigenvectors of real matrix. The difference between them is that the power method can calculate the main eigenvalues, which are the largest eigenvalues of matrix by modulo and its modulus is the spectral radius of the matrix and the corresponding eigenvectors, and inverse power method can calculate the smallest eigenvalues of the nonsingular (invertible) matrices by modulo and the corresponding eigenvectors. Now, power method and inverse power method are introduced as follows, respectively.

1. Power method (also known as multiplication power method)
   Suppose a real matrix $A = (a_{ij})$ has a complete set of feature vectors, namely A has n linear independent eigenvectors. Suppose the eigenvalues of matrix A are $\lambda_1, \lambda_2, \Lambda, \lambda_n$, then the corresponding eigenvectors are x_1, x_2, Λ, x_n, respectively. The main eigenvalue λ_1 of A is real root and meets the conditions: (1) $|\lambda_1| > |\lambda_2| \geq \cdots \geq |\lambda_n|$; (2) $Ax_i = \lambda_i x_i \quad (i = 1, 2, \ldots, n)$. Now we discuss how to get λ_1 and X_1.

   The basic idea of power method is to randomly select nonzero initial vector v_0 and construct a matrix vector sequence $\{v_k\}$ from matrix A:

$$\begin{cases} v_1 = Av_0 \\ v_2 = Av_1 = A^2v_0 \\ \vdots \\ v_{k+1} = Av_k = A^{k+1}v_0 \\ \vdots \end{cases} \tag{2.30}$$

known as the iteration vector, and v_0 can be uniquely represented as eq. (2.31) by the supposition.

$$v_0 = a_1x_1 + a_2x_2 + \cdots + a_nx_n \quad (\text{suppose } a_1 \neq 0). \tag{2.31}$$

Then

$$\begin{aligned} v_k = Av_{k-1} = A^kv_0 &= a_1\lambda_1^kx_1 + a_2\lambda_2^kx_2 + \cdots + a_n\lambda_n^kx_n \\ &= \lambda_1^k\left[a_1x_1 + \sum_{i=2}^{n} a_i(\lambda_i/\lambda_1)^kx_i\right] \equiv \lambda_1^k(a_1x_1 + \varepsilon_k), \end{aligned}$$

where

$$\varepsilon_k = \sum_{i=2}^{n} a_i(\lambda_i/\lambda_1)^k x_i.$$

From the assumption condition, we can find $\lim_{k\to\infty}\left(v_k/\lambda_1^k\right) = a_1x_1$ is the eigenvector of λ_1.

So when k is big enough, there is

$$v_k \approx \lambda_1^k a_1 x_1. \tag{2.32}$$

And it is an approximate eigenvector of the corresponding eigenvalue λ_1 of matrix A.

Due to

$$v_{k+1} = Av_k \approx \lambda_1^{k+1} a_1 x_1 \approx \lambda_1 v_k. \tag{2.33}$$

Using $(v_k)_i$ to represent the first i component of v_k, when the k is sufficiently large, there is

$$\frac{(v_{k+1})_i}{(v_k)_i} \approx \lambda_1, \tag{2.34}$$

which is the approximate value of the main eigenvalue λ_1 of A.

This method, which is according to nonzero vector V_0 and exponentiation A_K of matrix A to structure vector sequence $\{v_k\}$ to calculate the main eigenvalue λ_1 and the corresponding eigenvector (2.32), is called the power method.

Iterative formula is essentially constructed as the vector sequence $\{V_K\} = \{A_K V_0\}$ by multiplying the exponentiation A_k of matrix A and the nonzero vector V_0, then calculate principal eigenvalue λ_1 and the corresponding eigenvectors, which is the idea of the power method. The convergence rate of $(v_{k+1})_i/(v_k)_i \to \lambda_1 \quad (k \to \infty)$ is determined by the ratio of $r = |\lambda_2/\lambda_1|$, the smaller the r is, the faster the convergence rate is, but the convergence may be very slow when $r \approx 1$.

Theorem 2.8. *Suppose $A \in R_{n\times n}$ has n linearly independent eigenvectors and principal eigenvalue λ_1 satisfies $|\lambda_1| \geq |\lambda_2| \geq \Lambda \geq |\lambda_n|$, then for any nonzero initial vector $v_0 = u_0 (a_1 \neq 0)$, the power method is defined as*

$$\begin{cases} v_0 = u_0 \neq 0 \\ v_k = Au_{k-1} \\ \mu_k = \max(v_k) \quad (k = 1, 2, \ldots) \\ u_k = v_k/\mu_k \quad \textit{Vector standardization} \end{cases}, \tag{2.35}$$

then

$$\lim_{k\to\infty} u_k = \frac{x_1}{\max(x_1)}.$$

2. Inverse power method
Inverse power method can calculate the smallest eigenvalues of nonsingular matrix A according to the mode and the corresponding eigenvectors and can find any eigenvalue and corresponding eigenvector of matrix A when combining with the inverse power method of original point translation method.

Suppose matrix A is nonsingular, and its eigenvalue $\lambda_i(i = 1, 2, \Lambda, n)$ meets (1) $|\lambda_1| \geq |\lambda_2| \geq \cdots \geq |\lambda_{n-1}| > |\lambda_n| > 0$ and (2) $Ax_i = \lambda_i x_i \Leftrightarrow A^{-1}x_i = \lambda_i^{-1}x_i$, and its corresponding eigenvectors x_1, x_2, Λ, x_n are linearly independent, then the eigenvalue of A^{-1} is $1/\lambda_i$, and the corresponding eigenvector is still $x_i(i = 1, 2, \Lambda, n)$. At this point, the eigenvalue of A^{-1} meets

$$\left|\frac{1}{\lambda_n}\right| > \left|\frac{1}{\lambda_{n-1}}\right| \geq \cdots \geq \left|\frac{1}{\lambda_1}\right|.$$

Thus, the main eigenvalues $1/\lambda_n \approx \mu_k$ and eigenvectors $x_n \approx u_k$ when using power method to A and the smallest eigenvalues $1/\lambda_n \approx \mu_k A$ and the corresponding eigenvectors $x_n \approx u_k$. This method that solved A^{-1} is called inverse power method.

The iterative formula for the inverse power method is

$$\begin{cases} v_k = A^{-1}u_{k-1} \\ \mu_k = \max(v_k) \\ u_k = v_k/\mu_k \end{cases} \qquad (k = 1, 2, \ldots). \tag{2.36}$$

In order to avoid solving A^{-1}, v_k can be obtained by solving linear equations $Av_k = u_{k-1}$, and the LU decomposition method can be used, that is, the LU decomposition is first used, i.e., $A = LU$, the iterative formula for the inverse power method is

$$\begin{cases} \text{Solve } Lz_k = u_{k-1}, & \text{then find } z_k \\ \text{Solve } Uv_k = z_k, & \text{then find } v_k \\ \mu_k = \max(v_k) \\ u_k = v_k/\mu_k \end{cases} \qquad (k = 1, 2, \ldots), \tag{2.37}$$

where

$$\lambda_n \approx \frac{1}{\mu_k}, x_n \approx u_k.$$

For a given error ε, when $|\mu_k - \mu_{k-1}| < \varepsilon$, we get

$$\lambda_n \approx \frac{1}{\mu_k}, \quad x_n \approx u_k.$$

Obviously, the convergence rate of inverse power method depends on the ratio $|\lambda_n/\lambda_{n-1}|$; the smaller the ratio is, the faster the convergence is.

Theorem 2.9. *Let $A \in R_{n\times n}$ is nonsingular matrix and has n linearly independent eigenvectors, the eigenvalues satisfy $|\lambda_1| \geq |\lambda_2| \geq \Lambda \geq |\lambda_n - 2| > |\lambda_n| > 0$. Then, for any nonzero initial vectors $u_0(x_n \neq 0)$, vector sequence $\{v_k\}\{u_k\}$ constructed by the inverse power method must meet*

1. $\lim\limits_{k\to\infty} u_k = \frac{x_n}{\max(x_n)}$,
2. $\lim\limits_{k\to\infty} \max(v_k) = \frac{1}{\lambda_n}$.

In the inverse power method, original point translation method can also be used to accelerate the iterative process, or to get other eigenvalues of corresponding eigenvectors. If the matrix $(A - pI)^{-1}$ exists, its eigenvalues are

$$\frac{1}{\lambda_1 - p}, \frac{1}{\lambda_2 - p}, \ldots, \frac{1}{\lambda_n - p}. \tag{2.38}$$

The corresponding eigenvectors are always x_1, x_2, Λ, x_n. Now apply power method to the matrix $(A - pI)^{-1}$, and get the following iterative formula of the inverse power method:

$$\begin{cases} u_0 = v_0 \neq 0, & \text{Initial vector} \\ v_k = (A - pI)^{-1} u_{k-1} & \\ \vdots & \\ u_k = v_k / \max(\mu_k) & \end{cases} \quad (k = 1, 2, \ldots). \tag{2.39}$$

If p is an approximate value of the eigenvalue λ_j, which is separated from the other eigenvalues, that is

$$|\lambda_j - p| << |\lambda_i - p| \quad (i \neq j),$$

then $1/(\lambda_j - p)$ is the main eigenvalue of the matrix $(A - pI)^{-1}$, and the eigenvalues and eigenvectors can be calculated by the inverse power method, as shown in eq. (2.39). Suppose $A \in R_{n\times n}$ has n linearly independent eigenvectors x_1, x_2, Λ, x_n, then

$$u_0 = \sum_{i=1}^{n} a_i x_i \quad (a_j \neq 0), \tag{2.40}$$

$$v_k = \frac{(A - pI)^{-k} u_0}{\max((A - pI)^{-(k-1)} u_0)}, \tag{2.41}$$

$$u_k = \frac{(A - pI)^{-k} u_0}{\max((A - pI)^{-k} u_0)}, \tag{2.42}$$

where

$$(A - pI)^{-k} u_0 = \sum_{i=1}^{n} a_i (\lambda_i - p)^{-k} x_i.$$

Theorem 2.10. *Suppose $A \in R_{n\times n}$ has n linearly independent eigenvectors, the eigenvalues of matrix A and the corresponding eigenvectors are denoted as λ_i and $x_i (i = 1, 2, \ldots, n)$, and p is the approximation eigenvalue of λ_j, $(A - pI)^{-1}$ exists, and $|\lambda_j - p| << |\lambda_i - p| (i \neq j)$. Then, for any nonzero initial vector $u_0 (a_j \neq 0)$, the vector sequences $\{v_k\}$ and $\{u_k\}$ that constructed by the inverse power method (2.41) meets*

1. $\lim\limits_{k\to\infty} u_k = \frac{x_j}{\max(x_j)}$,
2. $\lim\limits_{k\to\infty} \max(v_k) = \frac{1}{\lambda_j - p}$, *i.e.*, $p + \frac{1}{\max(v_k)} \to \lambda_j$.

And the convergence rate is

$$r = |\lambda_j - p| \, / \, \min_{i\neq j} |\lambda_i - p| \, .$$

As known from the theorem, for $A - pI$ $(p \approx \lambda)$, the application of inverse power method can be used to calculate the eigenvector x_j, and only select p is the better approximation of λ_j and eigenvalue separation is better, the calculation of eigenvector can be finished with one or two iterations since r is very small.

In the inverse power method, the v_k is obtained by solving the equation group $(A - pI)v_k = u_{k-1}$; in order to save the workload, we can first decompose the $P(A - pI) = LU$, so v_k is compared with the solution of the two triangular equations

$$Ly_k = Pu_{k-1}, \qquad Uv_k = y_k.$$

The experiment shows that the selected u_0 is better according to the following method:

$$Uv_1 = L^{-1} P u_0 = (1, 1, \ldots, 1). \tag{2.43}$$

Using iteration method to solve eq. (2.43) and get v_1, formula (2.39) is iterated.

Theorem 2.11 (Singular Value Decomposition [SVD] Theorem). *Let A is the $m \times n$ complex matrix of rank $r (r > 0)$, then there exists unitary matrix U with order m and unitary matrix V with order n, which make*

$$(1)\ U^H AV = \begin{bmatrix} \Sigma & 0 \\ 0 & 0 \end{bmatrix},$$

where $\Sigma = diag(\sigma_1, \sigma_2, \ldots, \sigma_r)$ *and* $\sigma_i (i = 1, 2, \ldots, r)$ *are all nonzero singular values of the matrix* A.

Proof. Suppose the n eigenvalue of the Hermite matrix $A^H A$ is sorted by

$$\lambda_1 \geq \lambda_2 \geq \cdots \geq \lambda_r > \lambda_{r+1} = \cdots = \lambda_n = 0.$$

Then there is a unitary matrix V with order n, which makes

$$(2)\ V^H (A^H A) V = \begin{bmatrix} \lambda_1 & & \\ & \ddots & \\ & & \lambda_n \end{bmatrix} = \begin{bmatrix} \Sigma^2 & 0 \\ 0 & 0 \end{bmatrix}. \quad (2.44)$$

V is divided into $V = (V_1, V_2)$, where V_1 and V_2, respectively, are the first r columns and the final $n - r$ columns. And rewrite eq. (2.44) as

$$V^H A^H AV = V \begin{bmatrix} \Sigma^2 & 0 \\ 0 & 0 \end{bmatrix} V^H.$$

Then

$$A^H AV_1 = V_1 \Sigma^2 \ A^H AV_2 = 0. \quad (2.45)$$

From the first equation of Theorem 2.11, we can get

$$V_1^H A^H AV_1 = \Sigma^2 \quad \text{or} \quad (AV_1\Sigma)^H (AV_1\Sigma) = E_r.$$

From the second equation of Theorem 2.11, we can get

$$(AV_2)^H (AV_2) = 0 \quad \text{or} \quad AV_2 = 0.$$

Let $U_1 = AV_1\Sigma^{-1}$, then $U_1^H U_1 = E_r$, that is, the r columns of U_1 are orthogonal unit vectors to each other. Let $U_1 = (u_1, u_2, \ldots, u_r)$, then $u_1, u_2, \ldots, u_r$ can be expanded into the standard orthogonal basis of C^m, remark the added vector as $u_{r+1}, \ldots, u_m$, and structure matrix $U_2 = (u_{r+1}, \ldots, u_m)$, then

$$U = (U_1, U_2) = (u_1, u_2, \ldots, u_r, u_{r+1}, \ldots, u_m)$$

is a unitary matrix with order m, and $U_1^H U_1 = E_r$, $U_2^H U_1 = 0$.

So

$$U^H AV = U^H(AV_1, AV_2) = \begin{bmatrix} U_1^H \\ U_2^H \end{bmatrix} (U_1\Sigma, 0) = \begin{bmatrix} \Sigma & 0 \\ 0 & 0 \end{bmatrix}.$$

Thus

$$A = U \begin{bmatrix} \Sigma & 0 \\ 0 & 0 \end{bmatrix} V^H = \sigma_1 u_1 v_1^H + \sigma_2 u_2 v_2^H + \cdots + \sigma_r u_r v_r^H. \tag{2.46}$$

Equation (2.46) is called SVD of matrix A.

SVD of the matrix is not only widely applied in the aspects of linear equations, matrix norm, generalized inverse, optimization, and so on, but also applied in the fields of digital computing, digital image processing, information retrieval, psychology, and so on. Other interested readers can consult other related matrix books.

2.2.2 Schur Decomposition

In matrix decomposition, Schur decomposition is one of the common methods. This chapter mainly introduces some theorems about Schur decomposition in matrix theory [92].

Definition 2.4. Let $A, B \in R^{n\times n}(C^{n\times n})$, if one orthogonal (unitary) matrix U with order n makes

$$U^T AU = U^{-1}AU = B \quad (U^H AU = U^{-1}AU = B), \tag{2.47}$$

then it is called A orthogonal similar to B.

Theorem 2.12 (Schur Theorem). *Any n order complex matrix is similar to an upper triangular matrix, that is, there exists a matrix U with order n and an upper triangular matrix R with order n, which makes the*

$$U^H AU = R \tag{2.48}$$

where the diagonal element of R is the eigenvalue of A, which can be arranged according to the requirements.

Definition 2.5. Suppose $A \in C^{n\times n}$, if

$$AA^H = A^H A. \tag{2.49}$$

Then A is called the normal matrix.

Clearly, the diagonal matrix, Hermite matrix, anti-Hermite matrix, and orthogonal matrix are regular matrices.

Theorem 2.13 (Real Schur Decomposition). *Suppose $A \in R_{n\times n}$, there is an orthogonal matrix Q which makes*

$$Q^T AQ = \begin{pmatrix} R_{11} & R_{12} & \cdots & R_{1m} \\ & R_{22} & \cdots & R_{2m} \\ & & \ddots & \vdots \\ & & & R_{mm} \end{pmatrix}, \tag{2.50}$$

where $R_{ii}(i = 1, 2, \Lambda, n)$ is a first-order or second-order matrix, and each order R_{ii} is a real eigenvalue of A, and two eigenvalues of each second-order diagonal value are two conjugate complex eigenvalues of A.

2.2.3 QR Decomposition

Rutishauser proposed the LR algorithm that calculates the eigenvalues using the triangular decomposition of the matrix, and Francis built the QR algorithm that calculates the eigenvalues of the matrix using the QR decomposition of matrix. QR method is a transformation method, which is one of the most effective methods to compute the eigenvalues of a general matrix (small and medium matrices).

At present, the QR method is mainly used to calculate

1. the problem of calculating all eigenvalues of the upper Hessenberg matrix;
2. the problem of calculating all eigenvalues of symmetric triple-diagonal matrices.

Theorem 2.14 (Basic QR Method). *Suppose $A = A_1 \in R_{n\times n}$, and construct QR algorithm:*

$$\begin{cases} A_k = Q_k R_k & (\text{where } Q_k^T Q_k = I, \ R_k \text{ is upper triangular matrix}) \\ A_{k+1} = R_k Q_k & (k = 1, 2, \ldots) \end{cases} \tag{2.51}$$

and mark $\tilde{Q}_k = Q_1 Q_2 \cdots Q_k$, $\tilde{R}_k = R_k \cdots R_2 R_1$, then

1. *A_{k+1} is similar to A_k, namely $A_{k+1} = Q_k^T A Q_k$.*
2. *$A_{k+1} = (Q_1 Q_2 \cdots Q_k)^T A_1 (Q_1 Q_2 \cdots Q_k) = \tilde{Q}_k^T A_1 \tilde{Q}_k$.*
3. *$A^k = \tilde{Q}_k \tilde{R}_k$ is the decomposition equation of A^k.*

Theorem 2.15 (Convergence of QR Method). *Set $A = (a_{ij}) \in R^{n*n}$, if the eigenvalues of A meets: There are standard matrix $A_k, A = XDX^{-1}$ where $D = diag(\lambda_1, \lambda_2, \Lambda, \lambda_n)$, and suppose $X - 1$ has triangular decomposition $X - 1 = LU$ (L is an unit lower triangular matrix, U is an upper triangular matrix), then the $\{A_k\}$ generated by the QR algorithm essentially converges to an upper triangular matrix, that is*

$$A_k \overset{\text{In essence}}{\rightarrow} R = \begin{pmatrix} \lambda_1 & * & \cdots & * \\ & \lambda_2 & \cdots & * \\ & & \ddots & \vdots \\ & & & \lambda_n \end{pmatrix} \quad (\text{when } k \to \infty).$$

If we mark $A_K = ((a_{ij})_K)$, *then*

$$1.\ \lim_{k\to\infty} a_{ii}^{(k)} = \lambda_i, \tag{2.52}$$

$$2.\ \text{When } i > j,\ \lim_{k\to\infty} a_{ij}^{(k)} = 0. \tag{2.53}$$

Theorem 2.16. *If the symmetric matrix A meets the condition of Theorem 2.14, then* $\{A_K\}$ *generated by the QR algorithm converges to the diagonal matrix* $D = \text{diag}(\lambda_1, \lambda_2, \Lambda, \lambda_n)$.

Proof. The further results about convergence of QR algorithm can be known from Theorem 2.14: Let $A \in R_{n\times n}$ and A has complete collection of eigenvectors, if the eigenvalues of A have only real eigenvalues or multiple conjugate complex eigenvalues, $\{A_K\}$ produced by the QR algorithm converges to the sub-block upper triangular matrix (block diagonal for first-order and second-order block) and every 2×2 sub-block gives a pair of conjugate complex eigenvalue of A, and every first-order diagonal block is a real eigenvalue of A, namely

$$A_k \rightarrow \begin{pmatrix} \lambda_1 & \cdots & * & * & \cdots & * \\ & \ddots & \vdots & \vdots & & \vdots \\ & & \lambda_m & * & \cdots & * \\ & & & B_1 & \cdots & * \\ & & & & \ddots & \vdots \\ & & & & & B_l \end{pmatrix}. \tag{2.54}$$

where $m + 2l = n$, B_i is a 2×2 sub-block, which gives a pair of conjugate complex eigenvalues.

2.2.4 Hessenberg Decomposition

When we solve the problems of matrix eigenvalues, a simple process is to reduce the general real matrix A with elementary reflection matrix as orthogonal similarity reduction transformation to upper Hessenberg matrix. Thus, the problem of seeking eigenvalues of the original matrix is transformed to seek Hessenberg matrix.

Let $A \in R_{n\times n}$, here to explain how to choose the elementary reflection matrices $U_1, U_2, \Lambda, U_{n-2}$, and matrix A can reduce to an upper Hessenberg matrix by the orthogonal similarity transformation.

1. Let

$$A = \begin{pmatrix} a_{11} & a_{12} & \cdots & a_{1n} \\ a_{21} & a_{22} & \cdots & a_{2n} \\ \vdots & \vdots & & \vdots \\ a_{n1} & a_{n2} & \cdots & a_{nn} \end{pmatrix} = \begin{pmatrix} a_{11} & A_{12}^{(1)} \\ c_1 & A_{22}^{(1)} \end{pmatrix},$$

where $c_1 = (a_{21}, \Lambda, a_{n1})^T \in R^{n-1}$, let $c_1 \neq 0$, otherwise this step does not need to be reduced. So the elementary reflection matrix $R_1 = I - \beta_1^{-1} u_1 u_1^T$ can be chosen to make $R_1 c_1 = -\sigma_1 e_1$, where

$$\begin{cases} \sigma_1 = \operatorname{sgn}(a_{21}) \left(\sum_{i=2}^{n} a_{i1}^2 \right)^{1/2}, \\ u_1 = c_1 + \sigma_1 e_1, \\ \beta_1 = \sigma_1(\sigma_1 + a_{21}). \end{cases} \tag{2.55}$$

Let

$$U_1 = \begin{pmatrix} 1 & \\ & R_1 \end{pmatrix},$$

then

$$A_2 = U_1 A_1 U_1 = \begin{pmatrix} a_{11} & A_{12}^{(1)} R_1 \\ R_1 c_1 & R_1 A_{22}^{(1)} R_1 \end{pmatrix},$$

$$= \begin{pmatrix} a_{11} & a_{12}^{(2)} & a_{13}^{(2)} & \cdots & a_{1n}^{(2)} \\ -\sigma_1 & a_{22}^{(2)} & a_{23}^{(2)} & \cdots & a_{2n}^{(2)} \\ 0 & a_{32}^{(2)} & a_{33}^{(2)} & \cdots & a_{3n}^{(2)} \\ \vdots & \vdots & \vdots & & \vdots \\ 0 & a_{n2}^{(2)} & a_{n3}^{(2)} & \cdots & a_{nn}^{(2)} \end{pmatrix} = \begin{pmatrix} A_{11}^{(2)} & A_{12}^{(2)} \\ 0 \ c_1 & A_{22}^{(2)} \end{pmatrix},$$

where $c_1 = (a_{32}^{(2)}, \ldots, a_{n2}^{(2)})^T \in R^{n-2}$, $A_{22}^{(2)} \in R^{(n-2)\times(n-2)}$.

2. Step k reduction: repeat the above process, suppose A has completed step 1, step Λ, step $K-1$ orthogonal similarity transformation, then $A_k = U_{k-1} A_{k-1} U_{k-1}$, or

$$A_k = U_{k-1} \cdots U_1 A_1 U_1 \cdots U_{k-1},$$

and

$$A_k = \begin{pmatrix} a_{11}^{(1)} & a_{12}^{(2)} & \cdots & a_{1,k-1}^{(k-1)} & a_{1k}^{(k)} & a_{1,k+1}^{(k)} & \cdots & a_{1n}^{(k)} \\ -\sigma_1 & a_{22}^{(2)} & \cdots & a_{2,k-1}^{(k-1)} & a_{2k}^{(k)} & a_{2,k+1}^{(k)} & \cdots & a_{2n}^{(k)} \\ & & \ddots & \vdots & \vdots & \vdots & & \vdots \\ & & & -\sigma_{k-1} & a_{kk}^{(k)} & a_{k,k+1}^{(k)} & \cdots & a_{kn}^{(k)} \\ & & & & a_{k+1,k}^{(k)} & a_{k+1,k+1}^{(k)} & \cdots & a_{k+1,n}^{(k)} \\ & & & & \vdots & \vdots & & \vdots \\ & & & & a_{nk}^{(k)} & a_{n,k+1}^{(k)} & \cdots & a_{nn}^{(k)} \end{pmatrix}$$

$$\begin{matrix} k & n-k \end{matrix}$$

$$= \begin{pmatrix} A_{11}^{(k)} & A_{12}^{(k)} \\ 0\ c_k & A_{22}^{(k)} \end{pmatrix} \begin{matrix} k \\ n-k \end{matrix} ,$$

where $c_k = (a_{k+1,k}^{(k)}, \dots, a_{nk}^{(k)})^T \in R^{n-k}$, $A_{11}^{(k)}$ is the k-order upper Hessenberg matrix, $A_{22}^{(k)} \in R^{(n-k)\times(n-k)}$.

Let $c_k \neq 0$, then choose the elementary reflection matrix $c_k \neq 0$ to make $R_k c_k = -\sigma_k e_1$, where $c_k \neq 0$, then

$$\begin{cases} \sigma_k = \operatorname{sgn}\left(a_{k+1,k}^{(k)}\right)\left(\sum\limits_{i=k+1}^{n} (a_{ik}^{(k)})^2\right)^{1/2}, \\ u_k = c_k + \sigma_k e_1, \\ \beta_k = \sigma_k(\sigma_k + a_{k+1,k}^{(k)}), \\ R_k = I - \beta_k^{-1} u_k u_k^T. \end{cases} \tag{2.56}$$

Let $U_k = \begin{pmatrix} I & \\ & R_k \end{pmatrix}$, then

$$A_{k+1} = U_k A_k U_k = \begin{pmatrix} A_{11}^{(k+1)} & A_{12}^{(k)} R_k \\ 0\ R_k c_k & R_k A_{22}^{(k)} R_k \end{pmatrix} = \begin{pmatrix} A_{11}^{(k+1)} & A_{12}^{(k+1)} \\ 0\ c_{k+1} & A_{22}^{(k+1)} \end{pmatrix}, \tag{2.57}$$

where $A_{11}^{(k+1)}$ is the K+1-order upper Hessenberg matrix, and the first K step reduction only needs to calculate $A_{12}^{(k)} R_k$ and $R_k A_{22}^{(k)} R_k$ (when A is a symmetric matrix, we only need to calculate $R_k A_{22}^{(k)} R_k$).

3. Repeat the above process, then

$$U_{n-2} \cdots U_2 U_1 A U_1 U_2 \cdots U_{n-2} = \begin{pmatrix} a_{11} & * & * & \cdots & * & * \\ -\sigma_1 & a_{22}^{(2)} & * & \cdots & * & * \\ & -\sigma_2 & a_{33}^{(3)} & \cdots & * & * \\ & & \ddots & \ddots & \vdots & \vdots \\ & & & -\sigma_{n-2} & a_{n-1,n-1}^{(n-1)} & * \\ & & & & -\sigma_{n-1} & a_{nn}^{(n)} \end{pmatrix} = A_{n-1}.$$

Summing up the above-mentioned conclusions, Theorem 2.17 is obtained as follows:

Theorem 2.17 (Reduce Householder Matrix to the Upper Hessenberg Matrix). *Let $A \in R_{n\times n}$, then there is an elementary reflection matrix $U_1, U_2, \Lambda, U_{n-2}$ to make*

$$U_{n-2} \cdots U_2 U_1 A U_1 U_2 \cdots U_{n-2} = U_0^T A U_0 = H \tag{2.58}$$

as the upper Hessenberg matrix.

2.3 Conclusion

This chapter mainly discusses the common mathematical foundation used in digital watermarking, analyzes the common image transformations, and then introduces the common matrix decomposition. The results of image transform, matrix decomposition principle, technique, and the coefficient features of decomposed matrix will be used to research digital watermarking algorithm in the remaining chapters, and many mathematical theorems in this chapter will be through to the application of the whole book.

3 Color Image

This chapter mainly introduces some terms of color image and color space, and some basic image knowledge of color image and lays the foundation for subsequent studying of color image digital watermarking algorithm. Color image processing is a technique that makes the image to meet the requirements of the visual, psychology, and other aspects by analyzing, operating, and processing procedures. The previous color image processing was essentially limited to satellite images, while it has gained importance on the new possibilities in recent years. One of the reasons is that color image contains higher layers of information in relation with gray-level image, which allows color image processing to succeed in areas where the classical gray-level image processing dominates.

3.1 Introduction

In our daily life, human visions and actions are influenced by a lot of geometries and colorful information. When crossing a street, a technical apparatus is identified by its geometry as a traffic light. However, subsequently, only by analyzing color information can decide whether to continue, if the light is green, or stop, if the light is red. A camera-assisted driving information system should be able to evaluate similar information and either passes the information on to the driver of a vehicle or directly influence the behavior of the vehicle. The latter is important, such as the guidance of an autonomous vehicle on a public road. Something similar happen to the traffic signs, which can be classified as prohibitive, regulatory, or informative signs based on color and geometry.

The judgment of color information also plays an important role in individual object identification. We usually do not recognize friends or classmates from their appearances at a distance, usually we spot the color of the clothes and then affirm whether they are friends and classmates or not. The same thing applies to recognize an automobile in the parking lot. Generally, we don't search for model *X* of company *Y*, but rather we look for a red car, for example, only when a red vehicle is found, we judge whether that vehicle is the one we are looking or not. The searching strategy is driven by a hierarchical combination of colors and forms and is also applied in automatic object recognition systems [93].

3.2 The Basic Types of Image

Computer images generally use two methods to store static images: one is bit mapped, also called bitmap storage mode, the other is vector processing, also known as the vector storage mode. Bitmap describes graphic image in the form of field array, while vector graphs use mathematical methods to describe the geometry elements of graphic

DOI 10.1515/9783110487732-003

images. On the premise of enough amounts of files, bitmap file can truly reflect the level and color of the image, while the disadvantage is the larger file size, and it's suitable to describe pictures. The characteristic of vector class file is small capacities and can zoom without changing any image quality, which is suitable for describing graphs.

The map storage mode of bitmap is turning each pixel of the image to a data, then storing in a one- or two-dimensional matrix in bytes. For example, when the image is monochrome, a byte can store eight pixels of image data; for 16-color image, every two pixels are stored in a byte; for 256-color image, every pixel is stored in a byte. By that analogy, the bitmap is able to describe accurately various color modes of image. So the bitmap file is more suitable to store complex images and real photos (as the carrier of information hiding, bitmaps meet the most basic requirements), but it also has some drawbacks: with the improvement of resolution and color depth, disk space of bitmap images will increase sharply; at the same time, in the process of magnified image, the image will become fuzzy and distort. The storage patterns of vector just store the outline of the image, not every bit of pixel. For example, for a circular pattern, storing only the coordinate position of the center, the length of the radius, and the color of the internal and circular edge is enough. The drawback of this storage method is that it often spends a lot of time to do some complicated analysis of calculation work, but as for the image zooming, it does not affect the display precision, namely the image will not be anamorphic. And the storage space of image is much less than bitmap file. So vector processing is suitable for storing various charts and engineering design [94].

Our study of this chapter is mainly for the bitmap. As mentioned above, the bitmap can convert a pixel into a data and then store in a two-dimensional matrix; we defined three basic types of the image based on the existence style of the image palette and the corresponding relationship between numerical matrix and pixel color: binary image, gray-level image, and color image. Their performances are, respectively, introduced as follows.

3.2.1 Binary Image

Binary image is also known as black-and-white image or monochromatic image, usually using 1 or 0 to stand for black and white pixels, respectively, also known as a binary map. Each pixel values will be decided within discrete values (0 or 1), one refers to black, another stands for white. Binary images can be stored by using unsigned 8-bit integers (uint8) or double types of arrays.

Array of uint8 is usually better than double-type array, because uint8 array occupies much smaller memory. In MATLAB image processing toolbox, any function that returns a binary image uses uint8 logic array to store the image. Figure 3.1 shows the binary image of Lena.

Figure 3.1: The binary Lena image.

3.2.2 Gray-Level Image

Gray-level image is the image that includes gray level (brightness). Gray, in other words, is brightness. Different from binary image, gray-level image is still black and white in the sensory, but in fact it's not simple pure black (0) and white (1). Hence, every pixel is absolutely cannot be characterized by one bit.

In MATLAB, gray-level image is described by an array of uint8, uint16, or double precision. In fact, gray-level image is a data matrix I, and each entry of the matrix corresponds to a pixel of the image, the value of the element represents gray level in certain extent. Generally, 0 stands for black, and 1, 255, or 65535 (different store mode) refers to white.

The types of data matrix I may be double precision, uint8, or uint16. When storing gray-level image, the palette is not used, so MATLAB will use a default system palette to display the image. Gray-level image is different from black-and-white image, black-and-white image only have black and white colors in the field of computer image; and there are many levels of color depth in gray-level image between black and white. But, out of the field of digital image, the "black-and-white image" also represents "gray-level image," so the binary image can be seen as a special case of gray-level image. Linking with the YCbCr color space that would be mentioned later, we can find that the so-called gray-level image pixel value is the value of each pixel luminance component in YCbCr. Both have the same conversion relationships with RGB pixel. Figure 3.2 shows the gray-level image of Lena.

Figure 3.2: The gray-level Lena image.

3.2.3 Color Image

Intuitively, a color image is the perceived representation of our colorful surrounding environment (i.e., to one's individual visual sensory perception). This kind of image does not use single palette, the color of each pixel is decided by the red, green, and blue color components which stores in the corresponding position. Generally, achromatic color means white, black, and various degrees of gray color; achromatic color image is also called the gray-level image, which uses the values of [0, 255] to represent the gray value, 0 means black while 255 means white. A variety of different shades of gray in between, the pixels of whole image can be represented in a one-dimensional array. However, chromatic color refers to the various colors without the black, white, and gray, and RGB model is one of the color models. Computationally, a color image is seen as a vector function (generally with three components). The range of the image function is a vector space, provided with a norm that is also called a color space. For a (three-channel) digital color image, three vector components u_1, u_2, u_3 are given for one image pixel (x,y):

$$C(x,y) = \left[u_1(x,y), u_2(x,y), u_3(x,y)\right]^T = [u_1, u_2, u_3]^T . \tag{3.1}$$

The colors represented by concrete value combinations of the vector components u_1, u_2, u_3 only have relative meanings. Each of the vectors $[u_1, u_2, u_3]^T$ with the generally integer components $0 \leq u_1, u_2, u_3 \leq G_{\max}$ characterizes a color in the basic color space. Example of color spaces are the RGB color space that is used to represent a color image on a monitor (additive color mixture), or the CMYK color space, which is used for printing a color image (subtractive color mixture).

Let's take an RGB image as an example to enhance the understanding of color images. An RGB image is 24 bit, where red, green, and blue components, respectively, take up eight bits. Theoretically, it contains 16M kinds of different colors, whose accuracies can represent the real color image.

In MATLAB, an RGB image contains uint8, uint16, or double types of arrays (often called RGB array) where the m and n stand for width and height of the image, respectively.

In a double type of RGB data set, each color component is a value within the scope of [0, 1], color component to (0, 0, 0) pixels will be displayed as black, while color component to (1, 1, 1) pixels will be displayed as white. Three color components of each pixel are stored in in the third dimension of the data set. For example, red, green, and blue components from pixel (10, 5) are stored in the RGB (10, 5, 1), RGB (10, 5, 2), and RGB (10, 5, 3), respectively.

3.3 The Basic Terminology of Color Images

At present, the terms of gray-level image processing have reached a consensus [95]. Neither of the color edge and the difference in color images is defined yet, nor what is the contrast of color image. There are a lot of technical terms are used in different ways, sometimes these terms are not accurate. The commonly basic terms in color image processing would be given in the following sections.

3.3.1 Color Edge

In gray-level image, edge refers to the discontinuity of the gray-level image while the color edge has not been clearly defined for color image. Several different definitions have been proposed on color edges.

A very old definition points out that the edge in the color image is the edge in the brightness image [96]. This definition ignores discontinuities in tonality or saturability. For example, two equally light objects of various colors are arranged in juxtaposition in a color image, and then the edges between the object geometries cannot be determined with this way. Since color image contains more information than gray-level image, more edge information is expected from color edge detection in general. However, this definition delivers no new information in relation to gray-value edge detection.

A second definition for a color edge states that an edge exists in the color image if at least one of the color components contains an edge. In this monochromatic-based definition, no new edge detection procedures are necessary. This presents the problem of accuracy of the localization of edges in the individual color channels. If the edges in the color channels are detected as being shifted by one pixel, then the merging of the results produces very wide edges. It cannot be easily determined which edge position in the image is the correct one given in this situation.

A third monochromatic-based definition for color edge is based on the calculation of the sum of absolute values of the gradients for the gradients that exceed a threshold value [97]. The results of the color edge detection by the two previously named definitions depend heavily on the basic color spaces. An image pixel that, for example, is identified in one color space as an edge point must not eventually be identified in another color space as an edge point (and vice versa).

All previously named definitions ignore the relationship between the vector components. Since a color image represents a vector-valued function, a discontinuity of chromatic information should also be defined in a vector-valued way. The fourth definition for a color edge can be obtained by the derivative, described in the previous section, of a (as a rule in digital color image processing three channel) color image. For a color pixel or color vector $C(x, y) = [u_1, u_2, \ldots, u_n]^T$, the variation of the image function at position (x, y) is described by the equation $\Delta C(x, y) = J\Delta(x, y)$. The direction along which the largest change or discontinuity in the chromatic image function is represented in the image by the eigenvector $J^T J$ corresponding to the largest eigenvalue. If the size of the change exceeds a certain value, then this is a sign for the existence of a color edge pixel.

A color edge pixel can also be defined with vector ordering statistics or vector-valued probability distribution function.

3.3.2 Derivative of Color Image

For a color component or a gray-level image $E(x, y)$, the gradient or the gradient vector is given by

$$\operatorname{grad}(E) = \left[\frac{\partial E}{\partial x}, \frac{\partial E}{\partial y}\right]^T = (E_x, E_y)^T. \tag{3.2}$$

Here, the indexes x and y are introduced as abbreviations that indicate the respective partial derivative of the function, that is, it holds

$$E_x = \frac{\partial E}{\partial x} \text{ and } E_y = \frac{\partial E}{\partial y}.$$

The absolute value of the gradient, as defined in eq. (3.3), is a measurement for the "height change" of the gray-level image function. It takes on the extreme value of zero for a constant gray-level plateau (in the ideal case $E(x, y)$ is a const number):

$$\left\| \operatorname{grad}(E) = \sqrt{\left(\frac{\partial E}{\partial x}\right)^2 + \left(\frac{\partial E}{\partial y}\right)^2} \right\|. \tag{3.3}$$

A three-channel color image can be described by a function $C: Z^2 \rightarrow Z^3$. This definition can be easily applied to n-channel color images. The differential of function C is given in matrix form by the functional matrix or Jacobian matrix J, which contains the first partial derivatives for each vector component. For a color vector in a color space with $C(x, y) = [u_1, u_2, u_3]$, the derivative is described at a location (x, y) by the equation $\Delta C(x, y) = J\Delta(x, y)$. It holds

$$J = \begin{bmatrix} \frac{\partial u_1}{\partial x} & \frac{\partial u_1}{\partial y} \\ \frac{\partial u_2}{\partial x} & \frac{\partial u_2}{\partial y} \\ \frac{\partial u_3}{\partial x} & \frac{\partial u_3}{\partial y} \end{bmatrix} = \begin{bmatrix} \operatorname{grad}(u_1) \\ \operatorname{grad}(u_2) \\ \operatorname{grad}(u_3) \end{bmatrix} = \begin{bmatrix} u_{1x} & u_{1y} \\ u_{2x} & u_{2y} \\ u_{3x} & u_{3y} \end{bmatrix} = (C_x, C_y), \tag{3.4}$$

where both vectors are indicated with C_x and C_y:

$$C_x = [u_{1x}, u_{2x}, u_{3x}]^T \text{ and } C_y = [u_{1y}, u_{2y}, u_{3y}]^T.$$

3.3.3 Contrast of Color Image

The term contrast is used ambiguously in the references. In the following, several examples (without claiming completeness) are introduced:

3.3.3.1 Relative Brightness Contrast

Contrast describes the relation between the brightness values in an image or section of an image. As measurement for the size of the contrast, for example, the Michelson contrast $[(I_{max} - I_{min})/(I_{max} + I_{min})]$ is used [98], where the largest appearing brightness value is indicated by I_{max} and the smallest appearing brightness value is denoted by I_{min}.

3.3.3.2 Simultaneous Brightness Contrast

The perceptual phenomenon of brightness perception of a surface in dependence on the lightness of the background is likewise indicated as contrast. For the illustration of this phenomenon, a gray surface surrounded by a white surface and a gray surface of the same lightness surrounded by a black surface is considered. The gray-on-white background is perceived as somewhat darker than the gray-on-black background [98]. An example is given in Figure 3.3.

3.3.3.3 Relative Saturation Contrast

The relation between the saturation values in a color image can be described as relative saturation contrast. In a color image with low brightness contrast, details can be distinguished from the background on the basis of differing color saturation.

3.3.3.4 Simultaneous Color Contrast

The detection of a colored surface depends likewise on the color of the surface surrounding it. A gray surface surrounded by a red ring appears, for example, bluish-green [99]. For the description of induced color, influenced by the color of the surrounding surface, the opponent color model is frequently implemented [100]. Davidoff describes the effect of color contrast as the change of color constancy in a systematic manner [101].

(a)

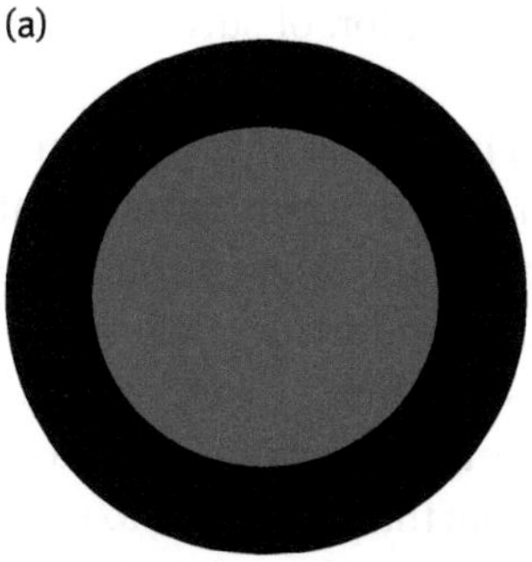

(b)

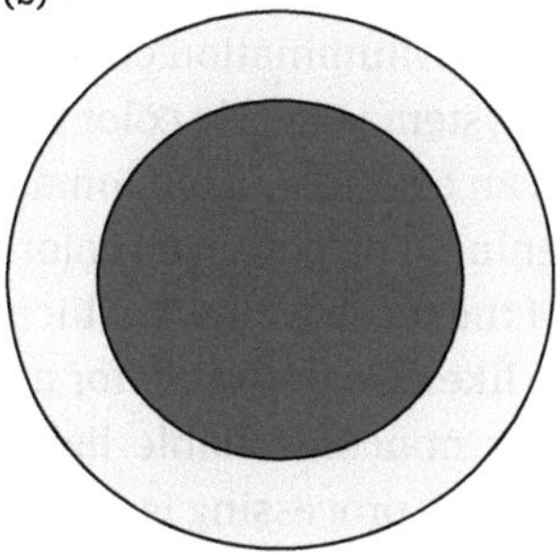

Figure 3.3: The example of brightness contrast: The gray-level circle in (a) is lighter than the one in (b).

3.3.3.5 Successive (Color) Contrast

This occurs when a colored area is observed over a long period of time and then a neutral area is subsequently fixed. An afterimage of the previously observed area appears either in the opponent colors (negative afterimage) or approximately in the previously observed colors (positive afterimage). Afterimages appear also with closed eyes.

Apart from the contrast definitions named here, the question is posed for digital color image processing as to what should be affected by the computer-aided change of contrast of a color image. The goal of enhancing the contrast in an image is generally to improve the visibility of image details. Only in rare cases the goal of the technique systematically influences the color constancy.

In many technical-based books, the contrast of a color image is regarded solely as brightness contrast in the sense of above [102]. Most display devices have implemented this definition for contrast control. On a color monitor (or television), the (nonlinear) area between the darkest and lightest pixels is adjusted with the "contrast control." With the "lightness control," a positive or negative offset for the lightness is established according to the adjustment. Also in the image-editing software program Adobe Photoshop, the function of contrast change corresponds to the lightness values of the image.

Color image processing offers the opportunity of changing the relative brightness contrast as well as the possibility of including perception-based observations if the need arises. In addition, color attributes such as saturation and intensity can also be set in relation to each other in the vector-valued color signals. A fact to be remembered is that the term contrast of a color image should be used with an adjective (e.g., relative or simultaneous) or an appropriate definition of the term.

3.3.4 Color Constancy

The colors of the surfaces of an object represent important features that could be used for identifying the object. However, a change in lighting characteristics can also change the several features of the light reflected from the object surfaces to the sensor. Color constancy is the capability of an invariant color classification of surfaces from color images with regard to illumination changes.

The human visual system is nearly color constant for a large area of surfaces and lighting conditions. As an example, a red tomato appears red in the early morning, at midday, and in the evening. The perceived color is therefore not the direct result of the spectral distribution of the received light, which was the assumption for many years.

Color constancy is likewise desirable for a camera-based vision system when its use should occur under noncontrollable lighting conditions. Achieving color constancy in digital color image processing is, however, a problem that is difficult to solve since the color signal measured with a camera depends not only on the spectral distribution of the illumination and the light reflected on the surface but also on the object geometry. These characteristics of the scene are, as a rule, unknown. In digital image processing, various techniques are identified for the numerically technical realization of color constancy. Color constancy techniques (in digital color image processing) can be classified into three classes with regard to the results that they intend to obtain:

1. The spectral distribution of the reflected light is to be estimated for each visible surface in the scene.
2. A color image of the acquired scene is to generate in the way it would appear under known lighting conditions.
3. Features are to be detected for the colored object surfaces in the image that are independent from lighting conditions (invariant to illumination changes).

3.3.5 Noise in Color Image

In the process of image formation, collection, and transmission, it will be affected by a certain degree of noise interference due to the existence of various interference factors. These interferences deteriorate the quality of the image and will affect all aspects of image processing and output. Hence, we should research the color image noise.

It is generally assumed that the individual components of the vector-valued color signal are degraded separately from each other by noise and that not all components are equally affected. This can be described, for example, by various additive overlays of the signals in the individual color components by malfunctions or Gaussian noise. Here the following model is used as a basis:

$$y = x + n, \tag{3.5}$$

where x denotes the undisturbed image vector at a position (i, j) in the color image. The corresponding vector with noise is indicated by y and n is an additive noise vector at position (i, j) in the image.

It cannot be concluded from the assumption of the existence of differing overlays in the individual color components that monochromatic-based techniques for separate noise suppression in the individual color components provide the best results. Vector-valued techniques allow, in general, a better treatment of noise in color image [103–105].

3.3.6 Luminance, Illuminance, and Brightness of Color Image

The terms luminance, illuminance, and brightness are often misused in color image processing. To clarify the terminology, we could apply three definitions from Adelson [106]:

Luminance is the amount of visible light that comes to the eye from a surface. In other words, it is the amount of visible light that leaves a point on a surface in a given direction due to reflection, transmission, and/or emission. Photometric brightness is an old and deprecated term for luminance. The standard unit of luminance is candela per square meter (cd/m^2), which is also called *nit* in the United States, and it is "to shine" in Latin nitere (1 nit = 1 cd/m^2).

Illuminance is the amount of light incident on a surface. It is the total amount of visible light illuminating (incident upon) a point on a surface from all directions above the surface. Therefore, illuminance is equivalent to irradiance weighted with the response curve of the human eye. The standard unit for illuminance is lux (lx), which is lumens per square meter (lm/m^2).

Brightness is the perceived intensity of light coming from the image itself, rather than any properties of the portrayed scene. Brightness is sometimes defined as perceived luminance.

3.4 Common Color Space of Color Image

Color is a perceived phenomenon from visual instead of a physical dimension like length or temperature, although the electromagnetic radiation of the visible wavelength spectrum is measurable as a physical quantity. The observer can perceive two differing color sensation wholly as equal or as metameric. Color identification through data of a spectrum is not useful for labeling colors. In image processing, a suitable form of representation must be found for storing, displaying, and processing color images. This representation must be well suited to the technical conditions of color image processing, and to human color perception as well. These various demands cannot be met equally well simultaneously. For this reason, differing representations are used in color image processing according to the processing goal.

3.4.1 RGB Color Space

The most commonly employed color space in computer technology is the RGB color space, According to the principle of colorimetry, all sorts of color lights in the nature can be composed by red, green, blue colors of light by different mixture ratio, and also the nature of all sorts of color of light can be decomposed into red, green, and blue colors of light. Hence, red, green, and blue colors are referred to as the three primary colors. Therefore, RGB color space is based on the additive mixture of three primary colors R, G, and B. The international standardized wavelengths of the primary colors red, green, and blue were already given in Table 3.1. It should be noted that the terms red, green, and blue were introduced only for the purpose of standardization to provide descriptions for the primary colors. Visible colors and wavelengths are not equivalent. In order to avoid possible confusion, the notations L, M, S may be used for light containing long, middle, and short wavelengths instead of the notations R, G, B. However, the usual notations are R, G, and B and they will be used in the following section.

The primary colors are for the most part the "reference colors" of the imaging sensors. They form the base vectors of a three-dimensional orthogonal (color) vector

Table 3.1: Wavelengths and the corresponding relative spectral power *S* of the CIE 1931.

Primary	Λ (nm)	*S*
R	700.0	72.09
G	546.1	1.379
B	435.8	1.000

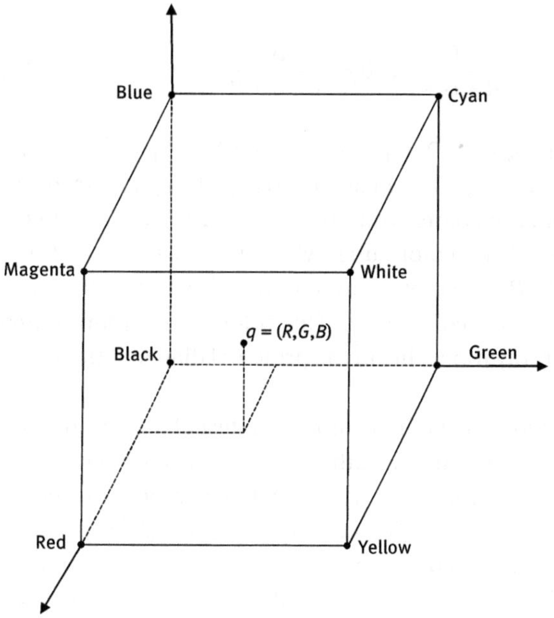

Figure 3.4: The color cube in the RGB color space.

space, where the zero vector represents black (see Figure 3.4). The origin is also described as black point. Any color can therefore be viewed as a linear combination of the base vectors in the RGB space. In one such accepted RGB color space, a color image is mathematically treated as a vector function with three components. The three vector components are determined by the measured intensities of visible light in the long-wave, middle-wave, and short-wave area. For a (three-channel) digital color image C, three vector components R, G, and B are to be indicated for such each image pixel (x,y):

$$C(x,y) = \left[R(x,y), G(x,y), B(x,y)\right]^T = (R,G,B)^T. \tag{3.6}$$

In the RGB color space, every vector of color cube precisely represents a color, $0 \le R$, G, $B \le G_{\text{max}}$, R, G, and B are integers. These values are referred to as tristimulus values. The colors that are represented by explicit value combinations of the vector components R, G, B are relative, device-dependent entities. All vectors with integer components $0 \leq R, G, B \leq G_{max}$ characterize one color in the RGB color space. $G_{max} + 1$ indicates the largest permitted value in each vector component. Using permeable filters in the generation of a color image in the RGB color space, so-called red, green, and blue extracts are generated in the long wave, middle wave, and short wave areas of visible light. If one refrains from using the filter, each of the three scannings is identical with the digitalization of a gray-level image. The following rational numbers r, g, b in eq. (3.7) are the color value components that are normalized with respect to the intensity:

$$r = \frac{R}{R+G+B}, \quad g = \frac{G}{R+G+B}, \quad b = \frac{B}{R+G+B}. \tag{3.7}$$

The primary colors red $(G_{max}, 0, 0)^T$, green $(0, G_{max}, 0)^T$, blue $(0, 0, G_{max})^T$, and the complementary colors yellow $(G_{max}, G_{max}, 0)^T$, magenta $(G_{max}, 0, G_{max})^T$, cyan $(0, G_{max}, G_{max})^T$, as well as the achromatic colors white $(G_{max}, G_{max}, G_{max})^T$ and black $(0, 0, 0)^T$ represent the boundaries of the color cube, which is formed through the possible value combinations of R, G, B. All color vectors $(R, G, B)^T$ with $0 \leq R, G, B \leq G_{max}$ each characterize a color in the RGB color space. This color cube is represented in Figure 3.4. All achromatic colors (gray tones) lie on the principal diagonal $(u, u, u)^T$, with $0 \leq u \leq G_{max}$.

The RGB color space is the most used in computer internal representation of color images. Its wide distribution is, among other things, traced back to the well-standardized three primary colors. Almost all visible colors can be represented by a linear combination of the three vectors. For identical objects, differing color values are generated with different cameras or scanners since their primary colors in general do not match with each other. The process of adjusting color values between different devices (e.g., camera RGB, monitor RGB, and printer RGB) is called color management [107, 108].

A special case of the RGB color space is the primary color system $R_NG_NB_N$ for television receivers (receiver primary color system), which refers to the established phosphors in the American standard NTSC (National Television System Committee). Values deviating from this are valid for the phosphors through the television standards PAL (phase alternation line) and SECAM (sequential color storage). The RGB color space, which was determined by CIE, is transformed into the NTSC primary color system $R_NG_NB_N$ [97]:

$$\begin{bmatrix} R_N \\ G_N \\ B_N \end{bmatrix} = \begin{bmatrix} 0.842 & 0.156 & 0.091 \\ -0.129 & 1.320 & -0.203 \\ 0.008 & -0.069 & 0.897 \end{bmatrix} \bullet \begin{bmatrix} R \\ G \\ B \end{bmatrix}. \tag{3.8}$$

3.4.2 YIQ Color Space

In the development of the NTSC television system used in the United States, a color coordinate system with the coordinates Y, I, and Q was defined for transmission purpose. In the YIQ system, the component Y refers to the brightness information of image, and components I, Q refer to color information of image, that is, component I stands for the color changing from orange to cyan while the component Q stands for the color changing from purple to yellow-green.

In order to transmit a color signal efficiently, the $R_NG_NB_N$ signal was more conveniently coded from a linear transformation. The luminance signal is coded in the Y component. The additional portions I (in-phase) and Q (quadrature) contain the entire chromaticity information that is also denoted as chrominance signal in television technology.

I and Q are transmitted by a much narrow waveband since the Y signal contains by far the largest part of the information. The Y signal contains no color information so that the YIQ system remains compatible with the black-white system. By using only the Y signal in a black-and-white television, gray-level images can be displayed, which would not be possible by a direct transmission of the $R_NG_NB_N$ signal.

The values in the $R_NG_NB_N$ color space can be transformed into the values in the YIQ color space by the following equation:

$$\begin{bmatrix} Y \\ I \\ Q \end{bmatrix} = \begin{bmatrix} 0.299 & 0.587 & 0.144 \\ 0.596 & -0.274 & -0.322 \\ 0.211 & -0.523 & 0.312 \end{bmatrix} \bullet \begin{bmatrix} R_N \\ G_N \\ B_N \end{bmatrix}. \tag{3.9}$$

3.4.3 YUV Color Space

The color television systems PAL and SECAM, developed in Germany and France, use the YUV color space for transmission. The Y component is identical with the one of the YIQ color space. The values in the $R_NG_NB_N$ color space can be transformed into the values in the YUV color space with eq. (3.10) [97]:

$$\begin{bmatrix} Y \\ U \\ V \end{bmatrix} = \begin{bmatrix} 0.299 & 0.587 & 0.144 \\ 0.418 & -0.289 & 0.437 \\ 0.615 & -0.515 & -0.100 \end{bmatrix} \bullet \begin{bmatrix} R_N \\ G_N \\ B_N \end{bmatrix} \tag{3.10}$$

On account of the low information content, the U and V signals, which are usually related to the Y signal, are reduced by half (two successive image pixels each having a separate Y portion, but with a common color type) and by a quarter for simple demands.

The I and Q signals of the YIQ color space are determined from the U and V signals of the YUV color space by a simple rotation in the color coordinate system, that is,

$$I = -U\sin(33°) + V\cos(33°),$$
$$Q = U\cos(33°) + V\sin(33°).$$

The presentations in the YIQ and YUV color space are very suitable for image compression since luminance and chrominance can be coded with different number of bits, which is not possible when using RGB values.

In the literature, YUV also indicates a color space, in which U corresponds to the color difference red-blue and V to the color difference green-magenta. Y corresponds to the equally weighted (arithmetical) averages of red, green, and blue. This color space is, for example, employed in highlight analysis of color images. We will denote this color space with (YUV)′ for a better distinction. A linear correlation exists between the (YUV)′ color space and the RGB system, which is given by the following transformation:

$$(Y, U, V)' = (R, G, B)\begin{bmatrix} \frac{1}{3} & \frac{1}{2} & \frac{-1}{2\sqrt{3}} \\ \frac{1}{3} & 0 & \frac{1}{\sqrt{3}} \\ \frac{1}{3} & \frac{-1}{2} & \frac{-1}{2\sqrt{3}} \end{bmatrix}. \tag{3.11}$$

Brightness normalization can be defined by

$$u = \frac{U}{R+G+B} \text{ and } v = \frac{V}{R+G+B}. \tag{3.12}$$

If u and v form the axes of a Cartesian coordinate system, then red, green, and blue stretch an equilateral triangle in which black lies in the origin of Figure 3.5.

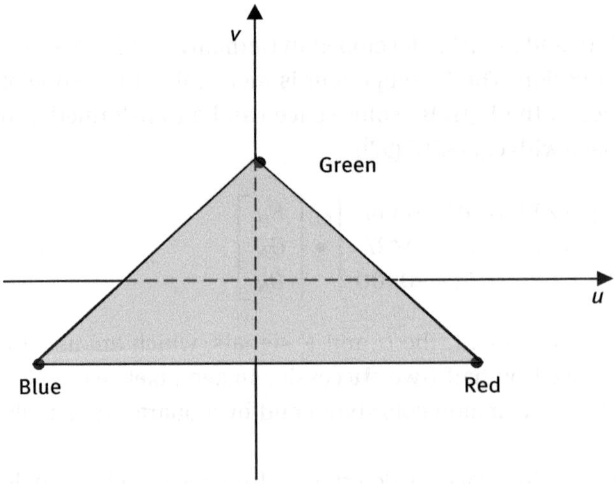

Figure 3.5: The uv-plane of the $(YUV)'$ model.

3.4.4 YCbCr Color Space

In the area of digital video, which is increasingly gaining importance, the internationally standardized YCbCr color space is employed for the representation of color vectors. This color space differs from the color space used in analog video recording. The values in the $R_N G_N B_N$ color space can be transformed into the values in the YCbCr color space [102]:

$$\begin{bmatrix} Y \\ C_b \\ C_r \end{bmatrix} = \begin{bmatrix} 16 \\ 128 \\ 128 \end{bmatrix} + \frac{1}{256} \begin{bmatrix} 65.738 & 129.057 & 25.064 \\ -37.945 & -74.494 & 112.439 \\ 112.439 & -94.154 & -18.285 \end{bmatrix} \bullet \begin{bmatrix} R_N \\ G_N \\ B_N \end{bmatrix}. \tag{3.13}$$

From this transformation it is assumed that the RGB data has already undergone a gamma correction. The quantities for the Y components refer to the fixed values for phosphors in the reference Rec. ITU-R BT. 601-4 of the NTSC system. The back-transformation from the YCbCr color space into the $R_N G_N B_N$ color space is (except for a few rounding errors) given by [102]

$$\begin{bmatrix} R_N \\ G_N \\ B_N \end{bmatrix} = \frac{1}{256} \begin{bmatrix} 298.082 & 0.0 & 408.583 \\ 298.082 & -100.291 & -208.120 \\ 298.082 & 516.411 & 0.0 \end{bmatrix} \bullet \begin{bmatrix} Y - 16 \\ C_b - 128 \\ C_r - 128 \end{bmatrix}. \tag{3.14}$$

The YCbCr color space was developed for representations in the television format common until now. It isn't applied to the high-definition television format.

3.5 Perception-Based Color Spaces

Color spaces that are based intuitively on human color perception are of importance in the fields of computer vision and computer graphics. In the HIS and the HSV color spaces, the wish for user-friendly input and description of color values is first considered. A color can be more easily described intuitively (above all by untrained users) by values for hue, color saturation, and intensity than from vector components in the RGB or CMYK color space.

3.5.1 HSI Color Space

In the HIS color space, hue, saturation, and intensity are used as coordinate axes.

Hue represents color, and color is related to the wavelength of color light. In order of red, orange, yellow, green, blue, purple color to defined tonal value and angle value is used to represent it. For example, angle value of red, yellow, green, blue, blue, magenta is 0°, 60°, 120°, 180°, 240°, and 300°, respectively.

Saturation represents the purity of color, that is, the degree of mixed white in color light. The more the white light is, the lower the saturation is, and the less the white light is, the higher the saturation is and the purer the color is. Saturation values adopt percentage (0–100%): 0% represents gray light or white light and 100% refers to the pure color.

Strength refers to the degree of light color sensed by human eye; it is related to the size and color of light energy (or the brightness of the light color), so the brightness is also sometimes used to present.

Usually the hue and saturation are collectively known as chroma, which is used to represent the category of the color and the degree of depth. The human visual system is more sensitive to the brightness of the degree than the sensitive degree of color shade, compared with the RGB color space, and the characteristics of human visual system use the HSI color space to explain more suitable. HSI color description is of man's natural, intuitive, which is in accord with human visual characteristic, and HSI model in the development of image processing method based on color description is also an ideal tool, for example, in HSI color space, it can be independently performed by algorithm of hue, saturation, and brightness. Adopting the HSI color space sometimes can reduce the complexity of the color image processing, improve the rapidity of processing, and enhance the understanding of color and explanation at the same time.

HSI color space is a conical space model, as shown in Figure 3.6; with this kind of cone model for describing that the HSI color space is quite complex, cone model's hue,

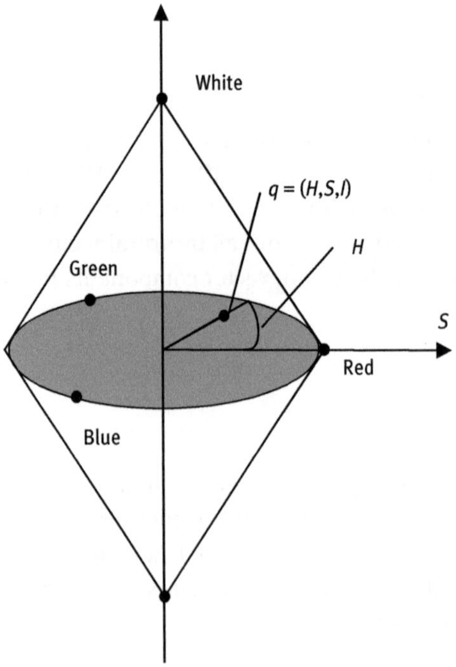

Figure 3.6: The HIS color space.

intensity, and the change of the relationship between the saturation can be clearly shown. Among them:

1. Line diagram: cone on brightness, chroma, and saturation.
2. The vertical axis represents brightness: brightness values are along the axis of the cone metric, along the point of the cone axis said not fully saturated color; according to different gray levels, the brightest spot is for the pure white and dark spot for pure black.
3. Cone longitudinal section: describe the relationship between the same hue of different brightness and saturation.
4. Cone section: tonal H is around the cone section measurement of color ring, the color on the circumference of a circle is completely saturated pure color, and color saturation is radius horizontal axis through the center.

This color space is well suited for the processing of color images and for visually defining interpretable local characteristics. A color $q = (R, G, B)^T$ is given in the RGB color space. The hue H of the color q characterizes the domain color contained in q. Red is specified as a "reference color." Because of that, $H = 0°$ and $H = 360°$ correspond to the color red. Formally, H is given by

$$H = \begin{cases} \delta & \text{if} \quad B \leq G, \\ 360° - \delta & \text{if} \quad B > G, \end{cases} \tag{3.15}$$

where

$$\delta = \arccos\left\{\frac{(R-G)+(R-B)}{2\sqrt{(R-G)^2+(R-B)(G-B)}}\right\}.$$

The saturation S of the color q is a measurement of color purity. This parameter is dependent on the number of wavelengths that contribute to the color perception. The wider the range of the wavelengths is, the lower the purity of the color is. The narrower the range of the wavelength is, the higher the purity of the color is. The extreme case $S = 1$ is true for a pure color and the extreme case $S = 0$ for an achromatic color. S is given by

$$S = 1 - 3\frac{\min(R, G, B)}{R + G + B}. \tag{3.16}$$

The intensity I of the color q corresponds to the relative brightness (in the sense of a gray-level image). The intensity is defined as follows:

$$I = \frac{R + G + B}{3}. \tag{3.17}$$

For the color $q = (R, G, B)^T$ in the RGB color space, a representation $(H, S, I)^T$ of this color is given in the HIS color space. This conversion is clearly reversible (except for inaccuracies in rounding and some singularities). The back-transformation is given below.

if ($S = 0$) then (gray tone)

$R = G = B = I;$

else

if ($0 \leq H \leq 120$) then (B is minimum)

$B = (1 - S) \cdot I;$

$H = \frac{1}{\sqrt{3}} \cdot \tan(H - 60);$

$G = (1.5 + 1.5 \cdot H) \cdot I - (0.5 + 1.5 \cdot H) \cdot B;$

$R = 3 \cdot I - G - B;$

else

if ($120 \leq H \leq 240$) then (R is minimum)

$R = (1 - S) \cdot I;$

$H = \frac{1}{\sqrt{3}} \cdot \tan(H - 180);$

$B = (1.5 + 1.5 \cdot H) \cdot I - (0.5 + 1.5 \cdot H) \cdot R;$

$G = 3 \cdot I - B - R;$

else (G is minimum)

$G = (1 - S) \cdot I;$

$H = \frac{1}{\sqrt{3}} \cdot \tan(H - 300);$

$R = (1.5 + 1.5 \cdot H) \cdot I - (0.5 + 1.5 \cdot H) \cdot G;$

$B = 3 \cdot I - G - R;$

end{if}

end{if}

end{if}

One of the advantages of the HIS color space is the separation of chromatic and achromatic information. The existence of singularities is a disadvantage for the HIS color space. Furthermore, it must be observed that the information content and the reliability of the calculation of hue and saturation depend on the luminosity [109]. For achromatic colors, neither hue nor saturation is defined. Generally, the nonlinearity characteristic of the cameras can affect the HIS conversion unfavorably.

Transformations between the color spaces can be significantly accelerated when using hardware. Image processing boards are available for PCs and workstations that transfer a video image (in NTSC or PAL format) or an RGB image into an HIS image in real time. Inversely, the back-transformation of HIS into the RGB color space can be derived from eqs (3.15) to (3.17).

3.5.2 HSV Color Space

The HSV color space, which is also called the HSB color space, is particularly common in the field of computer graphics. As in the HSI color space, hue, saturation, and brightness values are used as coordinate axes. A hexacone is obtained by projecting the RGB unit cube along the diagonals of white to black to form the topside of the

HSV pyramid. The hue H is indicated as an angle around the vertical axle. As in the HIS color space, red is determined with $H = 0°$, green with $H = 120°$, and blue with $H = 240°$. In the HSV color model, each color and its complementary color difference is 180°. Saturation S value is from 0 to 1, so the radius of cone top is 1. HSV color model representing the color of the domain is a subset of the CIE chromaticity diagram, and the color of the saturation is 100% in this model, the purity of which is generally less than 100%. In the cone vertex (origin), $V = 0$, and H and S are not defined, on behalf of the black. The cone top center $S = 0$, $V = 1$, and H is not defined, on behalf of the white. HSV color space corresponds to the painter color matching method. In general, painters get different shades of color from a pure color by changing color thick and color dark, join in a kind of pure color white to change color, add black to change the color depth, at the same time adding different proportions of white, black can obtain a variety of different hues. The cone is shown in Figure 3.7.

Here, H represents the color of phase angle whose value range is 0–360°. S represents color saturation, and S is a scale value, ranging from 0 to 1. It is expressed as the selected color purity and the color of the largest of the ratio of purity; generally speaking, S represents the purity of certain color, the larger the value of S, the purer the color; the smaller the value of S, the more grayer the color. V represents the brightness of color, ranging from 0 to 1. V is equal to 0 indicates on the bottom of the cone, which

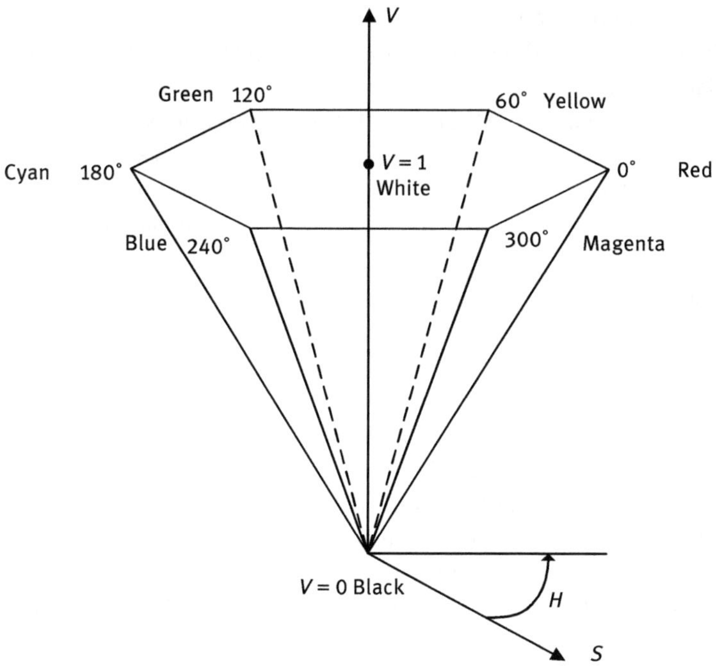

Figure 3.7: Hexacone representation of the HSV color space.

is black; V is equal to 1 in cone top surface; when $V = 1$ and $S = 0$ it is said to be pure white. However, not all colors with the same brightness are visible on the plane $V = 1$. The pseudocode for the conversion of a color image from the RGB color space into the HSV color space is indicated in the following, where again $G_{max} + 1$ denotes the largest possible value in each color channel. A representation of the back-transformation is not given here but can be found in Ref. [110]. Some image processing programs (e.g., Adobe Photoshop) contain modules for transformation images between the RGB and the HSV representation (called HSB).

Here is the conversion of color images from the RGB representation into an HSV representation:

```
max = Max(R, G, B);
min = Min(R, G, B);
V = max/Gmax;                                           {brightness value}
if(max ≠ 0) then S = (max − min)/max                    {saturation value}
else S = 0;
H = UNDEFINED
end{if}
if(S > 0)then
          DR = (max − R)/(max − min);
          DG = (max − G)/(max − min);
          DB = (max − B)/(max − min);
      if(max = R) then if(min = G) then H = 5 + DB      {color between magenta
                  else H = 1 − DG                           and yellow}
      else if(max = G) then if(min = B) then H = 1 + DR {color between yellow
      else H = 3 − DB                                       and cyan}
else if(min = R) then H = 3 + DG                        {color between cyan
else H = 5 − DR                                         and magenta}
end{if}
if(H < 6)then H = 60
else H = 0;
end{if}
```

With both the HSV and the HIS color spaces described in the previous paragraph, there exists the problem, apart from the singularities in the color space already specified, that a straight line in the RGB space is not generally mapped onto a straight line in the two other color models. Here in particular it is to be noted in the cases of interpolations in the color spaces and transformations between the color spaces. An advantage of the HSV color space lies in the fact that it intuitively corresponds to the color system of a painter when mixing the colors and its operation is very easy to learn. In digital color image processing, the HSV color space is of only secondary importance. It is used for the easily operated manipulation of a color image's color values (e.g., with Adobe Photoshop).

3.6 Conclusion

From our recognition and judgment on color information, first of all, this chapter introduces the basic types of image and basic terminology of color image, then introduces some commonly used color space and color space based on the perception above this basis, and lays a foundation to introduce the processing of the color image watermark. Color images contain higher information than gray-level image. The information of color image processing has been more widely used in the traditional image processing fields. For a lot of technologies, the adoption of color not only makes image processing process more simple, robust, and available, but also makes the study of color images watermarking more meaningful.

4 The Color Image Watermarking Algorithm Based on DC Component

In this chapter, a color image blind watermarking algorithm based on direct current (DC) coefficient in the spatial domain is proposed, which combines the advantages of better robustness in the frequency domain and lower complexity in the spatial domain. According to the formation principle of DC coefficient in discrete cosine transform (DCT) domain, each DC coefficient of 8×8 sub-block in luminance Y of host image and its modification value are obtained in spatial domain, then modifying DC coefficient in DCT domain to embed watermark by directly modifying pixel values; the original watermark and the original host image are not necessarily required during the extraction of watermark. The experimental results show that the proposed algorithm has better watermark performance.

4.1 Introduction

With the better awareness of digital copyright protection among people, the digital watermarking technique has paid more and more attention [111, 112]. In digital watermarking technique, it can be divided into spatial domain watermarking technique [113] and frequency domain watermarking technique [114–119]. The frequency domain watermarking technique is to transform image in the frequency domain and to embed watermark by modifying the coefficient, which has good robustness; while the spatial algorithm usually embeds watermark into unimportant bits of pixel, which has easy calculation and lower complexity. Since the spatial domain or the transform domain has different advantages, it has been widely applied in the watermarking technique. However, these applications are single formed either on frequency domain or spatial domain, that is, both advantages are not well combined to apply. Although Shih et al. [120] proposed a method that combines the frequency domain with the spatial domain, both advantages are not really used to embed watermark, and each domain is selected to perform under different conditions. In principle, the frequency domain watermarking method is to distribute the signal energy on all pixels of the spatial domain, which means we can directly update the pixel value in the spatial domain instead of the frequency domain.

Based on the above discussion, a color image watermark algorithm combined with the pros of both is proposed in this chapter. First, the original host image is converted from RGB color space into YCbCr color space, and component Y is divided into 8×8 pixels block; then, calculate DC coefficients of each components in spatial domain according to the formation principle of DC coefficients in DCT domain, obtaining the modification of each DC coefficient according to watermark information and quantifying step size; last, perform directly the embedding and extraction of watermark in spatial domain according to the features of DC coefficient modifications. In this

DOI 10.1515/9783110487732-004

chapter, the proposed algorithm performs the process of embedding watermark in spatial domain, which avoids errors during the formation of coefficient conversion in frequency domain. The experimental results show that the proposed algorithm can embed binary watermark information into the color host image, which not only has the advantage on better robustness in frequency domain, but has other advantages such as high operating efficiency on algorithm in spatial domain.

4.2 The Technique of Modifying DC Coefficient in Spatial Domain

DCT is a common technique in transform domain, whose transform results contain a DC coefficient and many alternative current (AC) coefficients. Since the disadvantages that take long time to obtain DC coefficients with cosine function are existed, the technique of obtaining DC coefficient in spatial domain is mainly introduced as follows.

4.2.1 Obtaining DC Coefficient in the Spatial Domain

Given an image $f(x, y)$ of size $M \times N$ ($x = 0, 1, 2, \ldots, M-1, y = 0, 1, 2, \ldots, N-1$), its DCT definition is given as follows:

$$C(u, v) = \alpha_u \alpha_v \sum_{x=0}^{M-1} \sum_{y=0}^{N-1} f(x, y) \cos \frac{\pi(2x+1)u}{2M} \cos \frac{\pi(2y+1)v}{2N}, \tag{4.1}$$

where

$$\alpha_u = \begin{cases} \sqrt{1/M}, & u = 0 \\ \sqrt{2/M}, & 1 \le u \le M-1 \end{cases}, \quad \alpha_v = \begin{cases} \sqrt{1/N}, & v = 0 \\ \sqrt{2/N}, & 1 \le v \le N-1 \end{cases}. \tag{4.2}$$

Similarly, its inverse DCT (IDCT) definition is given as follows:

$$f(x, y) = \sum_{u=0}^{M-1} \sum_{v=0}^{N-1} \alpha_u \alpha_v C(u, v) \cos \frac{\pi(2x+1)u}{2M} \cos \frac{\pi(2y+1)v}{2N}. \tag{4.3}$$

When $u = 0, v = 0$, then the DC coefficient $C(0, 0)$ in DCT domain is given as follows according to eq. (4.1):

$$C(0, 0) = \frac{1}{\sqrt{MN}} \sum_{x=0}^{M-1} \sum_{y=0}^{N-1} f(x, y). \tag{4.4}$$

As shown from eq. (4.4), the DC coefficient $C(0, 0)$ can be obtained through simple arithmetical operation in the spatial domain without complicated DCT transform, which can reduce the time of cosine or inverse cosine calculation.

4.2.2 The Feasibility of Embedding Watermark with DC Coefficient in Spatial Domain

In general, the procedure of embedding watermarking in the DCT domain is to add watermark information on the DCT coefficients, and then obtain the watermarked image via the IDCT. In the following section, the feasibility of embedding watermark with DC coefficient will be proved from the view of energy conversation.

Supposing a signal $E(i, j)$ is added on any one coefficient $C(i, j)$ after DCT transform, where $i = 0, 1, \ldots, M-1, j = 0, 1, \ldots, N-1$, then the coefficient $C(i, j)$ becomes $C(i, j)^*$, which shows that

$$C(i, j)^* = C(i, j) + E(i, j). \tag{4.5}$$

Calculating the modified image $f(x, y)^*$ with IDCT

$$\begin{aligned} f(x, y)^* &= \sum_{i=0}^{M-1}\sum_{j=0}^{N-1} \alpha_i\alpha_j C(i, j)^* \cos\frac{\pi(2x+1)i}{2M}\cos\frac{\pi(2y+1)j}{2N} \\ &= \sum_{i=0}^{M-1}\sum_{j=0}^{N-1} \alpha_i\alpha_j C(i, j) \cos\frac{\pi(2x+1)i}{2M}\cos\frac{\pi(2y+1)j}{2N} \\ &\qquad + \alpha_i\alpha_j E(i, j)\cos\frac{\pi(2x+1)i}{2M}\cos\frac{\pi(2y+1)j}{2N} \\ &= f(x, y) + e(i, j), \end{aligned} \tag{4.6}$$

where

$$e(i, j) = \alpha_i\alpha_j E(i, j)\cos\frac{\pi(2x+1)i}{2M}\cos\frac{\pi(2y+1)j}{2N}, \tag{4.7}$$

and $e(i, j)$ is the signal that is added to the (i, j) DCT coefficient of the (x, y) pixel block. The energy of all signals can be calculated by

$$E = \sum_{x=0}^{M-1}\sum_{y=0}^{N-1} e^2(x, y). \tag{4.8}$$

According to eqs (4.2) and (4.7), eq. (4.8) can be further deduced by

1. When $i = 0, j = 0$, then

$$E = \sum_{x=0}^{M-1}\sum_{y=0}^{N-1} e^2(i,j) = \sum_{x=0}^{M-1}\sum_{y=0}^{N-1} \frac{1}{MN} E^2(i,j) = E^2(i,j)\,. \tag{4.9}$$

2. When $i = 0, j \neq 0$, then

$$\begin{aligned} E &= \sum_{x=0}^{M-1}\sum_{y=0}^{N-1} e^2(i,j) = \sum_{x=0}^{M-1} \frac{1}{M} E^2(i,j) \sum_{y=0}^{N-1} \frac{2}{N}\cos^2\left[\frac{(2y+1)j\pi}{2N}\right] \\ &= E^2(i,j)\sum_{y=0}^{N-1} \frac{2}{N}\cos^2\left[\frac{(2y+1)j\pi}{2N}\right] = E^2(i,j)\,. \end{aligned} \tag{4.10}$$

3. When $i \neq 0, j = 0$, then

$$\begin{aligned} E &= \sum_{x=0}^{M-1}\sum_{y=0}^{N-1} e^2(i,j) = \sum_{x=0}^{M-1} \frac{2}{M}\cos^2\left[\frac{(2x+1)i\pi}{2N}\right] \sum_{y=0}^{N-1} \frac{1}{N} E^2(i,j) \\ &= E^2(i,j)\sum_{x=0}^{M-1} \frac{2}{M}\cos^2\left[\frac{(2x+1)i\pi}{2M}\right] = E^2(i,j)\,. \end{aligned} \tag{4.11}$$

4. When $i \neq 0, j \neq 0$, then

$$\begin{aligned} E &= \sum_{x=0}^{M-1}\sum_{y=0}^{N-1} e^2(i,j) \\ &= \sum_{x=0}^{M-1} \frac{2}{M}\cos^2\left[\frac{(2x+1)i\pi}{2M}\right] \sum_{y=0}^{N-1} \frac{2}{N}\cos^2\left[\frac{(2y+1)j\pi}{2N}\right] E^2(i,j) \\ &= E^2(i,j)\,. \end{aligned} \tag{4.12}$$

From the above-mentioned equations, the modification of coefficients in any positions in DCT domain is equal to the changed quantity in spatial domain after IDCT. It has no exception with DC which is used to embed watermark according to eq. (4.9).

4.2.3 Modifying DC Coefficient in Spatial Domain

Apart from a DC coefficient obtained during DCT, others are AC coefficients, so the inverse transform can be written as

$$f(x,y) = \frac{1}{\sqrt{MN}} C(0,0) + f(x,y)^{AC}, \tag{4.13}$$

where $f^{AC}(x,y)$ is the reconstructed image from the set of AC components.

Suppose the host image is represented by

$$f(x, y) = \{f_{i,j}(m, n), 0 \le i < M/b, 0 \le j < N/b, 0 \le m, n < b\}, \tag{4.14}$$

where M and N are the row and column size of the host image, and the host image is divided into $i \times j$ nonoverlapped blocks with size of $b \times b$. The indexes of both rows and columns for each block are represented by (i, j), and (m, n) is the pixel coordination in each block.

It is assumed that when embedding watermark W into the DC component of the (i, j)th block, the modified value of the DC component is denoted as $\Delta M_{i,j}$, the traditional process of embedding the watermark into the DC component of the (i, j)th nonoverlapped $b \times b$ block is given by

$$C_{i,j}(0, 0)^* = C_{i,j}(0, 0) + \Delta M_{i,j}, \tag{4.15}$$

$$f_{i,j}(m, n)^* = \frac{1}{b} C_{i,j}(0, 0)^* + f_{i,j}(m, n)^{\mathrm{AC}}, \tag{4.16}$$

where $C_{i,j}(0, 0)$ is the DC coefficient of the (i, j)th block, $C'_{i,j}(0, 0)$ is the modified DC coefficient with increment $\Delta M_{i,j}$, and $f'_{i,j}(m, n)$ is the watermarked image. Equation (4.16) can be further deduced according to eqs (4.14) and (4.15) as follows:

$$\begin{aligned} f_{i,j}(m, n)^* &= \frac{1}{b} C_{i,j}(0, 0)^* + f_{i,j}(m, n)^{AC} \\ &= \frac{1}{b}(C_{i,j}(0, 0) + \Delta M_{i,j}) + f_{i,j}(m, n)^{AC} \\ &= \frac{1}{b}\Delta M_{i,j} + f_{i,j}(m, n) \end{aligned} \tag{4.17}$$

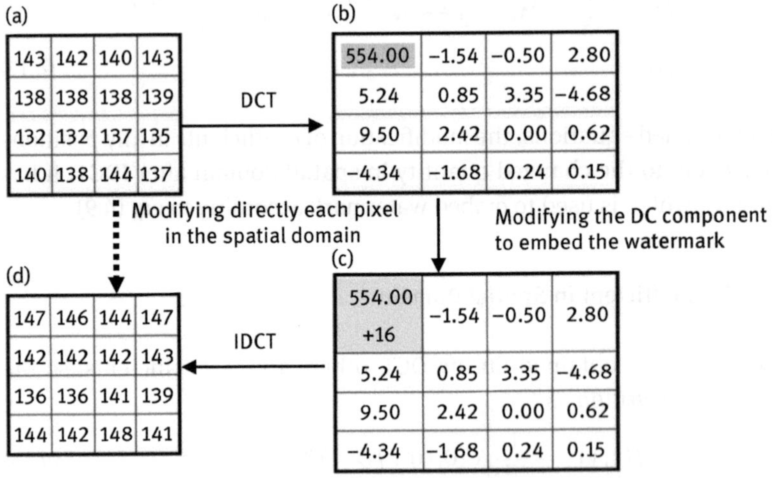

Figure 4.1: The illustration of DC embedding technique in spatial domain: (a) 4 × 4 pixel block, (b) DCT transformed block, (c) modifying DC component in the DCT domain, and (d) obtained by IDCT or eq. (4.17) in the spatial domain.

It is shown from eq. (4.17) that for the host image $f(x, y)$, the procedure of embedding the watermark into the DC component of DCT domain can also be directly performed in the spatial domain. That is, the modified value of each pixel in the spatial domain is $\Delta M_{i,j}/b$. In this chapter, an example of processing a block of size $b \times b$ is utilized to illustrate this process. The block of size 4×4 is shown in Figure 4.1(a). When embedding the watermark in DCT domain, the block via DCT transformation is displayed in Figure 4.1(b). Then the DC component is modified with $\Delta M = 16$. Figure 4.1(c) shows the embedded watermark according to eq. (4.15). Last, the watermarked block can be obtained by IDCT in eq. (4.3) and is given in Figure 4.1(d). Note that the difference of each corresponding pixel pair between Figure 4.1(a) and 4.1(d) is $\Delta M/b = 16/4 = 4$. That is, according to eq. (4.17), Figure 4.1(d) can also be directly obtained from Figure 4.1(a) in the spatial domain.

4.3 The Spatial Watermarking Algorithm Based on DC Coefficient

Generally, the procedure of watermarking includes watermark generation, watermark embedding, and watermark extraction. The Hash permuted algorithm is adopted to improve the security and robustness in the proposed method during the formation of watermark, and the quantization coefficient method is designed to embed watermark and can extract the embedded watermark with blind manner. The detailed algorithm is described in Section 4.3.1.

4.3.1 Watermark Generation

The original watermark W is employed in this chapter, which is shown in Figure 4.2(a), after Hash permutation based on key K_1 [128], and the original watermark image is rearranged to W as shown in Figure 4.2(b), which further improves the robustness and security of watermark.

4.3.2 Watermark Embedding

The detailed procedure of watermark embedding is, as shown in Figure 4.3, described as follows:

Figure 4.2: The preprocess of original watermark: (a) Original watermark and (b) permuted watermark.

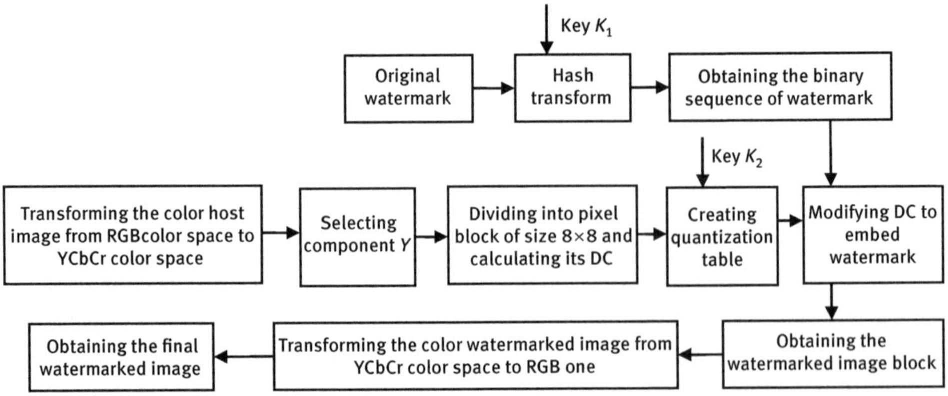

Figure 4.3: The diagram of watermark embedding.

1. Transform the host image from the RGB color space into the YCbCr color space.
2. Select the luminance Y of the YCbCr, and divide it into nonoverlapping 8×8 pixel blocks.
3. According to eq. (4.4), directly compute the DC coefficient $C_{i,j}(0,0)$ of each block in the spatial domain, where i and j represent the row and column indexes of each block, respectively.
4. According to $QA(k)$ and $QB(k)$, decide the DC coefficients, where Δ is the quantization step based on the secret key K_2:

$$QA(k) = \min(C_{i,j}(0,0)) + (2k - 4) \times \Delta, \tag{4.18}$$

$$QB(k) = \min(C_{i,j}(0,0)) + (2k - 5) \times \Delta, \tag{4.19}$$

   where $1 \leq k \leq \mathrm{round}((\max(C_{i,j}(0,0)) + 2\Delta)/(2\Delta)) - \mathrm{round}((\min(C_{i,j}(0,0)) - 2\Delta)/(2\Delta)))$, min(.), and max(.) denote the minimum value and maximum value of quantization coefficient, respectively, and round(.) is the module function.
5. Calculate the modified value $MC_{i,j}$ of the DC coefficient by eqs (4.20) and (4.21):

$$C_{i,j}(0,0)^* = \begin{cases} QA(k), & \text{if } W(i,j) = 1 \text{ and } \min(C_{i,j}(0,0) - QA(k)) \\ QB(k), & \text{if } W(i,j) = 0 \text{ and } \min(C_{i,j}(0,0) - QB(k)) \end{cases}, \tag{4.20}$$

$$MC_{i,j} = C_{i,j}(0,0)^* - C_{i,j}(0,0), \tag{4.21}$$

   where $C_{i,j}(0,0)^*$ is the DC coefficient after embedding watermark.
6. Add $MC_{i,j}/8$ to all pixels in the block by eq. (4.17), that is, embedding one watermark bit into this block in the spatial domain.
7. Repeat steps 3–6 until all blocks are performed to obtain the watermarked luminance Y. Then, transform the watermarked image from the YCbCr color space to RGB color space by obtaining the final watermarked image I^*.

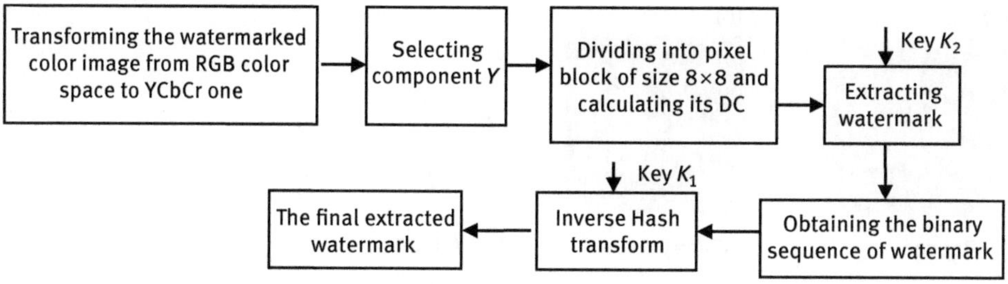

Figure 4.4: The diagram of watermark extraction.

4.3.3 Watermark Extraction

The extraction of the watermark without the requirement for the original image or the watermark image is shown in Figure 4.4, whose detailed steps are introduced as follows:

1. Transform the watermarked image I^* from the RGB color space into the RGB YCbCr color space.
2. Select the luminance Y of the YCbCr, and divide it into nonoverlapping 8×8 pixel blocks.
3. By eq. (4.4), directly obtain the DC coefficient $C_{i,j}$ (0,0).
4. Using the quantification step Δ based on the secret key K_2, compute the watermark $w'(i, j)$ as follows:

$$w(i,j)^* = \mathrm{mod}(\mathrm{ceil}(C_{i,j}(0,0)/\Delta), 2), \tag{4.22}$$

   where mod() is the module function and ceil(x) is the smallest integer which is not less than x.
5. Utilize the secret key K_1 to perform the inverse Hash transform on $w'(i, j)$ and obtain the extracted watermark image W'.

4.4 Algorithm Test and Result Analysis

In order to test the performance of the proposed algorithm, in this chapter, four 24-bits true color standard images with size of 512×512 are taken as the original host images, which are shown in Figure 4.5(a), and also the binary image size of 64×64 is used as digital watermark in Figure 4.2(a), which meets the requirements of the maximization of watermark length.

In order to solve the conflicts between watermark robustness and invisibility, the quantification step Δ is set to 20 according to JPEG's quantification matrix; in order to evaluate the quality of watermarked image, the structural similarity index measurement (SSIM) is employed to measure the similarity between watermarked

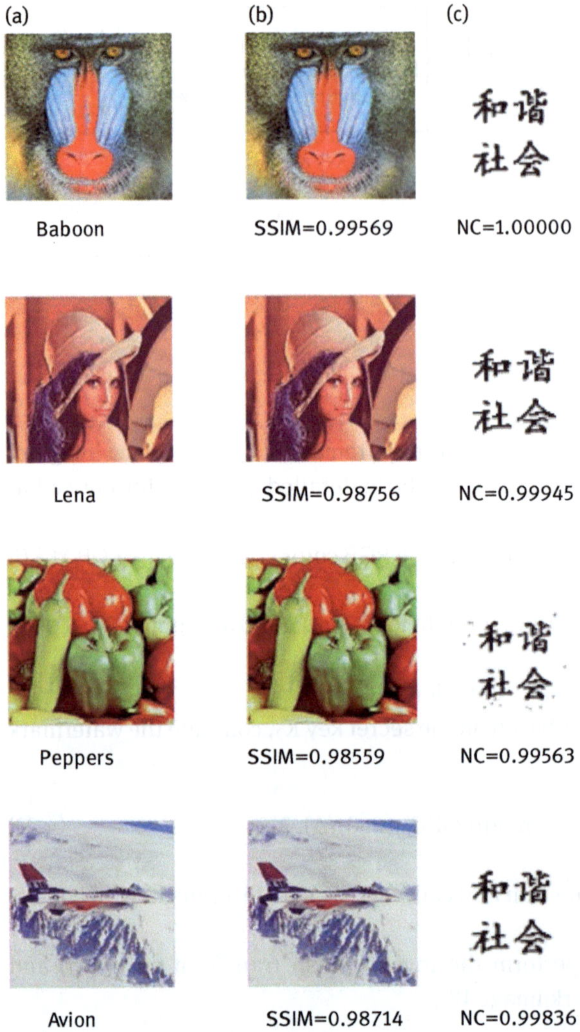

Figure 4.5: The results of embedding and extracting watermark without any attacks: (a) Original host image, (b) watermarked image, and (c) extracted watermark from (b) without any attacks.

image I^* and the original image I, and the normalized cross-correlation (NC) is used to measure the similarity between the extracted watermark W^* and the original watermark W.

4.4.1 Watermark Invisibility Test

Figure 4.5(a) is the original host image, Figure 4.5(b) is the corresponding watermarked image, and Figure 4.5(c) is the extracted watermark from Figure 4.5(b) without

Table 4.1: The comparison results of embedding watermark using DC in different domains and the results of the extracted watermark without any attacks.

Image	Spatial domain		DCT domain	
	SSIM	NC	SSIM	NC
Baboon	0.99569	1.00000	0.99568	0.98004
Lena	0.98756	0.99945	0.98755	0.95790
Peppers	0.98558	0.99563	0.98555	0.93576
Avion	0.98714	0.99836	0.98711	0.95435

any attacks. We can see from Figure 4.5(b) that the proposed algorithm has good watermark invisibility.

We can see from Table 4.1 that the proposed algorithm based on the spatial domain is superior to the algorithm based on DCT, this is because the algorithm based on DCT contains DCT and IDCT, which includes numeric-type conversion, cosine function calculation, matrix operation, and irrational number calculation. These calculation errors can result in lower calculation accuracy and larger differences.

Meanwhile, in this chapter, it obtains operating time data with different algorithms using Matlab 2010 experimental platform in hardware environment Pentium 2.80 GHz CPU and memory of 1 GB. From Table 4.2, it takes fewer time to operate in spatial domain than that in DCT domain, that is because the time complexity of the former is $O(N^2)$, while the time complexity of the latter is $O(N^2)$, so the proposed algorithm in the spatial domain is superior to the algorithm in DCT domain.

4.4.2 Watermark Robustness Test

To verify the robustness of the proposed watermarking, four watermarked images, as shown in Figure 4.5(b), are performed by the image processing operations (such as JPEG compression, adding noises, filtering, and mosaic attacks) and geometrical distortions (such as cropping operation).

Table 4.3 gives the results of four watermarked images under JPEG compression attacks; when the compression factor is 40 %, the value of NC is sufficient to show that the proposed watermark algorithm has good robustness.

Table 4.2: The comparisons of the performing time in different domains (seconds).

Time	Spatial domain	DCT domain
Embedding time	3.1044	5.3196
Extracting time	0.3901	1.8096
Total time	3.4945	7.2292

Table 4.3: The NC values of different watermarked images after JPEG compression attacks.

JPEG	Baboon	Lena	Peppers	Avion
30	0.8291	0.8201	0.7958	0.7359
40	0.9858	0.9639	0.9508	0.9527
50	0.9859	0.9806	0.9705	0.9790
60	0.9986	0.9970	0.9926	0.9885
70	0.9997	0.9995	0.9940	0.9986
80	0.9997	0.9992	0.9956	0.9995
90	1.0000	1.0000	0.9984	1.0000

Table 4.4: The NC values of different watermarked images after adding salt and pepper noise attacks.

Salt and peppers noise	Baboon	Lena	Peppers	Avion
0.01	0.9721	0.9785	0.9574	0.9709
0.02	0.9371	0.9428	0.9341	0.9303
0.03	0.9139	0.9423	0.8937	0.9090
0.04	0.8950	0.8931	0.8753	0.8707
0.05	0.8606	0.8712	0.8666	0.8486

Table 4.5: The NC values of different watermarked images after other attacks.

Attack	Baboon	Lena	Peppers	Avion
Mosaic 2 × 2	0.9984	0.9956	0.9863	0.9918
Mosaic 3 × 3	0.8158	0.8242	0.8242	0.8513
Median filtering 2 × 2	0.9978	0.9929	0.9917	0.9901
Median filtering 3 × 3	0.9648	0.9136	0.9475	0.8778
BLPF n = 2.50 Hz	0.8461	0.9075	0.9420	0.9017
BLPF n = 3.50 Hz	0.8261	0.9001	0.9410	0.9005

Table 4.4 is the result of image against different salt and pepper noises. We can see that all images still can extract watermark with larger NC values when adding different noise factors, which demonstrates the proposed algorithm has good robustness.

Meanwhile, Table 4.5 gives the results of watermarked images against different types of attacks, such as mosaic attack (2 × 2, 3 × 3), median filtering attack (2 × 2, 3 × 3), and Butterworth low-pass filtering attack (cutoff frequency is 50 Hz, class is n = 2, 3). It is shown in Figure 4.5 that the NC value of extracted watermark is almost near to 1, which means the proposed algorithm has good robustness against common attacks.

In order to further test the performance of proposed algorithm against geometric attacks, Figure 4.6(a) and 4.6(c) shows the results of watermarked Lena image against

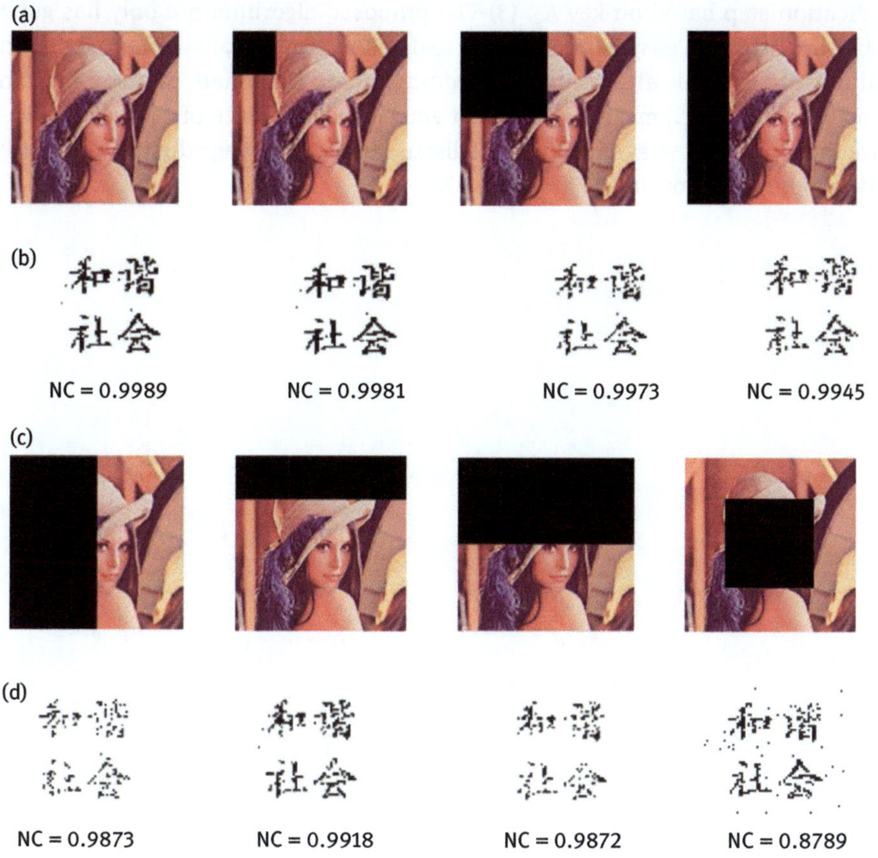

Figure 4.6: The results of extracted watermark under various cropping position to watermarked Lena image: (a) and (c) The cropped watermarked images; (c) and (d) extracted watermarks from (a) and (c).

cropping with different positions and sizes, Figure 4.6(b) and 4.6(d) give the extracted watermarks from corresponding cropping images, which shows that the proposed algorithm has good robustness according to the visual effects of extracted watermark and the values of NC.

4.5 Conclusion

In this chapter, a new blind color image is proposed, and its advantages are as follows: (1) It can perform the calculation of DC coefficients in DCT domain and can embed digital watermark into DC coefficients. Compared with other algorithms of DCT, the operating length of proposed algorithm reduces half of original time, lowers calculation errors, and improves the performance of algorithm. (2) It can enhance the security of watermark algorithm based on key K_1 Hash permutation transform and the quantification step based on key K_2. (3) The proposed algorithm not only has good robustness, but is also easy to operate and realize the purpose of blind extraction in spatial domain. But this algorithm is embedding binary watermark information into color host image, which means that it can't embed the same size of color image watermark into color host image, and we will discuss how color image digital watermark embed into the color host image in Chapter 5.

5 The Color Image Watermarking Algorithm Based on Integer Wavelet Transform

In this chapter, the state coding technique is proposed to make the state code of data set equal to the hiding watermark information by modifying a data in the data set. When embedding watermark, integer wavelet transform (IWT) and the rules of state coding are used to embed *R*, *G*, and *B* components of color image watermark into *R*, *G*, and *B* components of color host image. Moreover, the rules of state coding are also used to extract watermark from the watermarked image without the original watermark or the original host image. Experimental results show that the proposed watermarking algorithm can not only meet the requirements of the invisibility and robustness of watermark, but also have good watermark capacity, which is 98,304 bits.

5.1 Introduction

In recent years, the host images used in most of researched digital watermarking algorithms are gray-level images [36], while most watermarks are binary images [39, 64, 121] or gray-level images [40–42]. The watermarking algorithm that embeds binary images into color images is researched in the last chapter, which can't meet the requirement of embedding color watermark into color image, even though it can quickly protect the copyright of color image with binary image.

Since the color image, as digital watermark, contains more information, it is difficult to embed watermark into the host image compared with the binary watermark. Hence, the research on the color image watermarking is very few. While the color image is very popular on today's Internet, research on the embedding and extraction of color image watermark is more valuable than others.

For a feasible digital watermarking method, three critical technical measurements must be taken into consideration: watermark robustness, invisibility, and capacity [36]. For color image watermark, further improving the robustness is necessary under the premise of fully embedding watermark and keeping good invisibility. So how to secretly embed high-capacity color image watermark into host image is the first problem to be solved.

Since IWT has good characters such as anisotropy and diversify directivity, which can directly map one integer into another integer without round-off error, and can also operate quickly and provide good watermark transparency. Recent years, researchers proposed many digital watermarking algorithms based on IWT. Among current watermarking algorithms based on IWT [122–124], most of them focus on semi-fragile watermarking and their host images are gray, and only a few watermarking algorithms is applied to color image. For example, an image tamper detection algorithm based on the lifting IWT is proposed in Ref. [122]. A multiple marked watermarking

DOI 10.1515/9783110487732-005

method based on IWT for protecting the gray-level image copyright is proposed in Ref. [123]. In order to protect the copyright of color image, a blind watermark algorithm based on IWT, which embeds binary watermark into color image, is introduced in Ref. [124].

From the above discussion, a dual-color digital watermarking algorithm based on IWT and state coding is proposed in this chapter. On the one hand, the proposed state coding technology is used to make the state code of data set equal to the hidden watermark information, which not only guarantees the blind extraction of watermark, but increases the capacity of watermark. On the other hand, the advantage of IWT is used to enhance the robustness of watermark. The experimental results show that this algorithm not only meets the requirement of embedding color image watermark with high capacities, but also has good invisibility.

5.2 State Coding and IWT

5.2.1 State Coding Technique

In order to realize the blind extraction of watermark, we propose a new state coding method. Before that, some terms about state coding should be defined.

Definition 5.1. State code is the module between a value and its base number.

It is easy to calculate the state code of a single number, that is, the state code of decimal number 61 is 1; as for a data set, state code needs to be calculated by the following equation:

$$s = \text{mod}\left(\sum_{i=1}^{n} (a_i \times i), r\right), \tag{5.1}$$

where r refers to the base number of a_i. For example, the state code of a decimal data set {12, 34, 56} is $s = 8$, where the base number $r = 10$.

Definition 5.2. The method of state coding is a process that makes the state code of the data set equal to the hiding information by changing some of data in the data set.

In general, n integers may have $2n$ changed states since any integer x has two changed states, that is, increase or decrease. In this chapter, different change states represent different watermark information w, $w = \{0, 1, \ldots, 9\}$, and w includes 10 states; in addition, the state code of unit data formed by five decimal number can show one watermark information because decimal data has ten base-number codes, which is the same as the states of watermark.

5.2.2 Integer Wavelet Transform

Since color image digital watermark contains more data information, it takes long time to embed and detect, while traditional wavelet transform used by current watermarking algorithms will take long times to perform. Moreover, the outputs of traditional wavelet transform filter are float numbers, so there is round-off error with wavelet coefficients quantization, and the image's reconstruction quality is related with the way of boundary treatment; while the gray value of image is represented and stored as integer style, so undistorted transform is necessary. IWT can directly map integer into integer without round-off error, which can also be performed quickly [125, 126].

Laboratory AT&Bell WIM Sweldens proposed a lifting scheme that can be used in IWT [127]. The lifting scheme is a new method of constructing wavelets, which is not constructed by Fourier transform or based on Fourier transform scale contraction, but through a series of simple steps such as the split, predict, and update procedures to transform the digital signal, and its detailed steps are introduced as follows:

1. Split: The inputted original signal s_i is divided into two small mutually disjoint subsets, which includes odd subset s_{i-1} and even subset d_{i-1}, namely wavelet subsets. The simplest split method can be denoted as

$$F(s_i) = (s_{i-1}, d_{i-1}). \tag{5.2}$$

2. Predict: In general, these two subsets are closely related, and one subset can be well predicted with another subset. In practice, although it is not possible to accurately predict subset d_{i-1} from s_{i-1}, it can use the difference between d_{i-1} and $P(s_{i-1})$ to replace the original subset d_{i-1}, which contains less information than the original d_{i-1}, namely

$$d_{i-1} = d_{i-1} - P(s_{i-1}), \tag{5.3}$$

where P is the prediction operator, which needs to take original signal features into consideration and to reflect the correlations between data.

3. Update: In order to maintain some global characteristics of the original signal in the subset s_{i-1}, for example, to remain the entire image brightness value of the original image in the sub-image, the update operation must be performed to keep the original image and sub-image to have same average pixel brightness values. The target of update operation is to find a better subset s_{i-1} that keeps certain scalar property of the original image (e.g., mean and disappear moment invariant), and the update operation is defined as follows:

$$s_{i-1} = s_{i-1} + U(d_{i-1}), \tag{5.4}$$

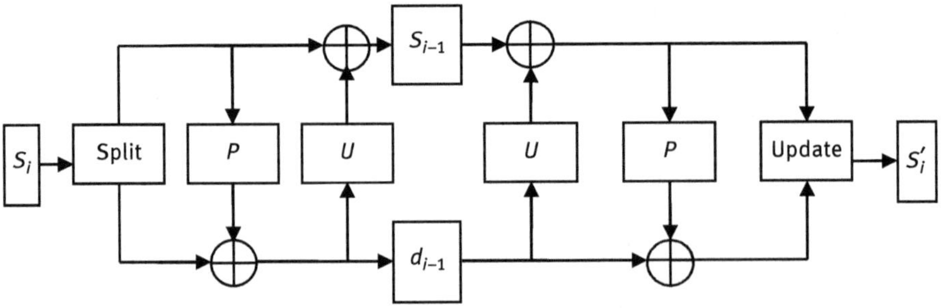

Figure 5.1: The decomposition and reconstruction of lifting scheme.

where U is the update operator, the subset d_{i-1} becomes the low-frequency component after the lifting transformation, and the odd subset s_{i-1} becomes the high-frequency component. The same transformation for the low-frequency components can get to the next level transformation. The decomposition and reconstruction of lifting scheme are shown in Figure 5.1.

5.3 The Color Image Watermarking Algorithm Based on State Coding and IWT

5.3.1 Watermark Embedding

The digital processing of color image used as digital watermark is very important before embedding watermark, and the structure of the digital watermark will directly affect the quality of embedded image, and the digital processing of color image is more complicated than the one of the binary image. Before embedding watermark, first, divide the color watermark image, which is used as digital watermark, into three components: red (R), green (G), and blue (B). In addition, in order to enhance the robustness of digital watermarking, the Hash permutation based on MD5 is applied to rearrange the pixel position of each component Wr, Wg, and Wb in this chapter [128]. On the basis of IWT, the state coding method is used to embed watermark in the proposed embedded algorithm, which is shown in Figure 5.2, and the specific watermark embedding steps are introduced as follows:

1. The integer watermark information Wr, Wg, and Wb are, respectively, processed to one-dimensional data by dimension-reduction treatment, and each pixel value is converted to the string with the length of three characters. Then, they are successively joined to form the final watermark information. For example, three pixel values, 206, 66, and 5 are converted to "206", "066", and "005," respectively, which form the character type data "206066005" to be used as watermark.
2. Divide the host color image I into R, G, and B components according to the three primary colors and perform one-level IWT to obtain the low-frequency coefficients Hr, Hg, and Hb of each component.

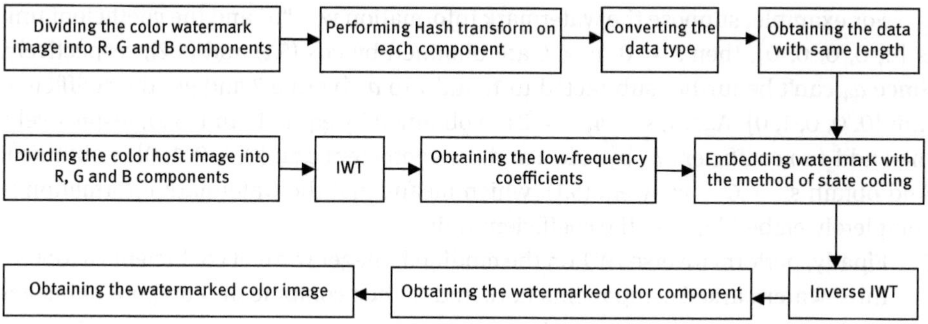

Figure 5.2: The diagram of watermark embedding.

3. Embed the components (Wr, Wg, and Wb) of watermark into the different positions of the components (Hr, Hg, and Hb) of host image with state coding.

The detailed steps of state coding are described as follows:
1. Separate all low-frequency coefficients of the host image into continual wavelet coefficient units, and each unit includes five coefficients, which can be represented as $\{a_i, 1 \le i \le 5, 0 \le a_i \le 255\}$
2. Use eq. (5.1) to calculate the state code s of the current coefficient unit.
3. Intercept the current watermark information w from the character type watermark information, and calculate the difference e between the w and the original state code s as follows:

$$e = \text{mod}((w - s), 10). \tag{5.5}$$

4. Embed watermark w according to the following rules until $s = w$.
   - Rule 1: if $e = 0$, then not to modify any value of the coefficient unit.
   - Rule 2: if $1 \le e \le 5$, then add 1 to a_e of the coefficient unit.
   - Rule 3: if $e > 5$, then subtract 1 from a_{10-e} of the coefficient unit.

For example, suppose the watermark information w is "1," and the coefficient unit is {137,139,141,140,130}. Thus, $s = 8$, $e = 3$ are obtained based on eqs (5.1) and (5.2), respectively. Moreover, add 1 to the third coefficient by rule 2, then the modified coefficient unit will be {137, 139, 142, 140, 130}.

Additionally, two special cases need to give attention.

Case 1: when $1 \le e \le 5$ and $a_e = 255$, use rule 2 to add 1 to a_e, which will make the a_e more than 255, which exceeds the valid range [0, 255]. At this time, the addition operation will be replaced by subtraction operation.

Case 2: when $e > 5$ and $a_{10-e} = 0$, use rule 3 to subtract 1 from a_{10-e}, which will make the a_{10-e} less than 0, which also exceeds the valid range [0, 255]. At this time, the subtraction operation will be replaced by addition operation.

For example, suppose the watermark information w = "6" and the coefficient unit is {0, 0, 0, 0, 0}, then $s = 0$, $e = 4$ are obtained by eqs (5.1) and (5.2), respectively. Since a_4 can't be further subtracted to 1, add 1 to a_4 by rule 2 and get the coefficient unit {0, 0, 0, 1, 0}. Again, $s' = 4$, $e' = 2$ are obtained by eqs (5.1) and (5.2), respectively, then add 1 to coefficient a_2 by rule 2 and the coefficient unit {0, 1, 0, 1, 0} is obtained. And obtain $s'' = 6 = w$ by eq. (5.1), which means that the watermark information is completely embedded into the coefficient unit.

5. Finally, perform inverse IWT on the modified integer wavelet coefficients to obtain three watermarked components R, G, and B, then combine the component images to obtain watermarked image I^*.

5.3.2 Watermark Extraction

The watermark extraction is easy, when the steps of embedding procedure are understood, and it is the inverse procedure of the embedding procedure. The procedure of watermark extraction is shown as Figure 5.3, and its detailed steps are introduced as follows:

1. Divide the watermarked color image I^* into three components R, G, and B according to the three primary colors, and perform the first IWT, then low-frequency coefficients Tr, Tg, and Tb are obtained.
2. Every coefficient unit is combined with five coefficient values from each low-frequency coefficients.
3. Extract watermark information from the coefficient unit according to eq. (5.1).
4. Three extracted watermark information is combined to a pixel value of watermark, and the final three watermark components R, G, B can be obtained, respectively.
5. Combine three watermark components into the final watermark W^*.

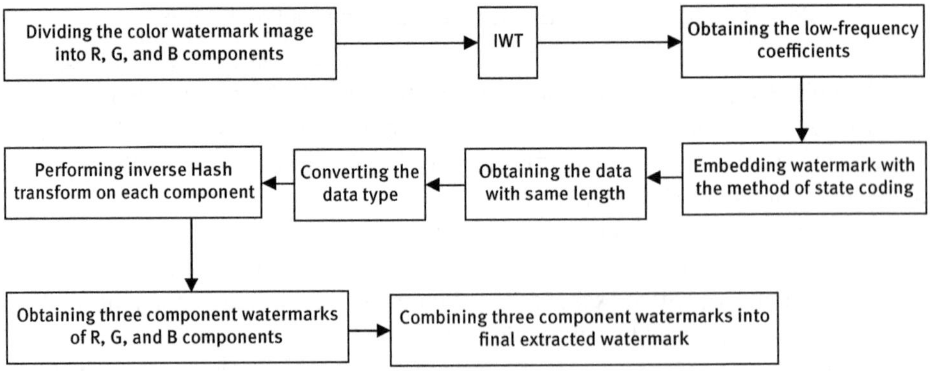

Figure 5.3: The diagram of watermark extraction.

5.4 Experimental Results and Discussion

In this chapter, four 24-bit standard color images with size of 512 × 512 are selected as the original host images for testing the performance of the proposed algorithm, which is shown in Figure 5.4(a) and one 24-bit color images with size of 64 × 64, as shown in Figure 5.4(b), is taken as the original color image watermark.

This chapter uses structural similarity index measurement (SSIM) to evaluate the similarity between original color host image I and watermarked color image I^*, namely to evaluate the invisibility of watermark. At the same time, normalized cross-correlation (NC) is also used as objective measurement criteria standard to measure between extracted watermark W^* and host watermark W, namely to evaluate the robustness of watermark.

5.4.1 Watermark Invisibility Test

From Figure 5.5(a), the SSIM value of extracted watermark image is near to 1, which makes difficult to notice the existence of embedded watermark, so the watermark method proposed in this chapter has good invisibility; meanwhile, the experimental results show that it is easy to extract the embedded watermark when the watermarked image is under no attacks, which is shown in Figure 5.5(b).

5.4.2 Watermark Robustness Test

JPEG compression attack is one of the common attacks that must be verified in watermarking algorithm, so it is very important for the proposed watermark algorithm to perform JPEG compression attack. In this experiment, JPEG compression attack with

Figure 5.4: Original image: (a) original host image and (b) original watermark image.

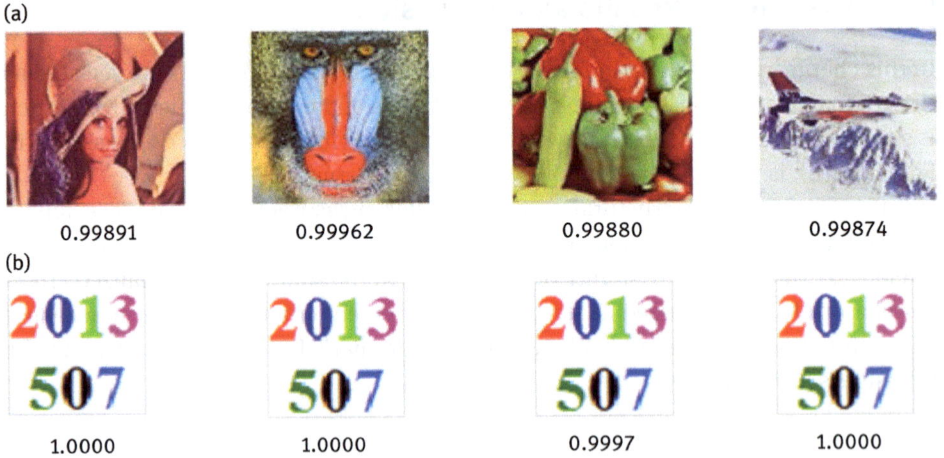

Figure 5.5: The results of embedding watermark and the results of the extracted watermark without any attacks: (a) Watermarked image (SSIM) and (b) extracted watermark from (a) (NC).

different compression factors from 10 to 100 on watermarked image is performed. It is clear that the smaller the compression factor is, the lower the quality of the extracted watermark is. Figure 5.6 gives the experimental results under different attacks. As shown in Figure 5.6(a), the proposed method has very good robustness against JPEG compression. Although the compression factor is reduced to 20, the NC value of watermark image is still valid; it represents the relevancy between the original watermark and extracted watermark. Generally, the watermark is valid if the NC is more than or equal to 0.750, conversely maybe invalid [129].

Figure 5.6(b) is the result after attacked by adding salt and peppers noise, whose NC values are close to 1, which explains this method has good robustness against attacking by adding salt and peppers noise.

The filtering attacking is also one of the common attacks. The spatial-based median filtering and the frequency-based Butterworth low-pass filtering are used to attack the watermarked image, and the extracted results are shown in Figure 5.6(c) and 5.6(d), respectively. From Figure 5.6(c), the proposed watermark has good robustness when the size of template is not more than 5×5. Figure 5.6(d) shows that the watermark can be extracted from different images with different blur radius when the watermarked image is attacked by 50 Hz cutoff frequency.

Figure 5.6(e) is the result of mosaic process, which shows that the bigger the mosaic size is, the lower the robustness quality is, and it has good robustness when mosaic size is less than 5×5.

The above-mentioned attacks are some common image processing attacks. Moreover, the geometric attacks, such as image rotation and image cropping, are adopted in this chapter. Figure 5.6(f) is the result from four watermarked images with

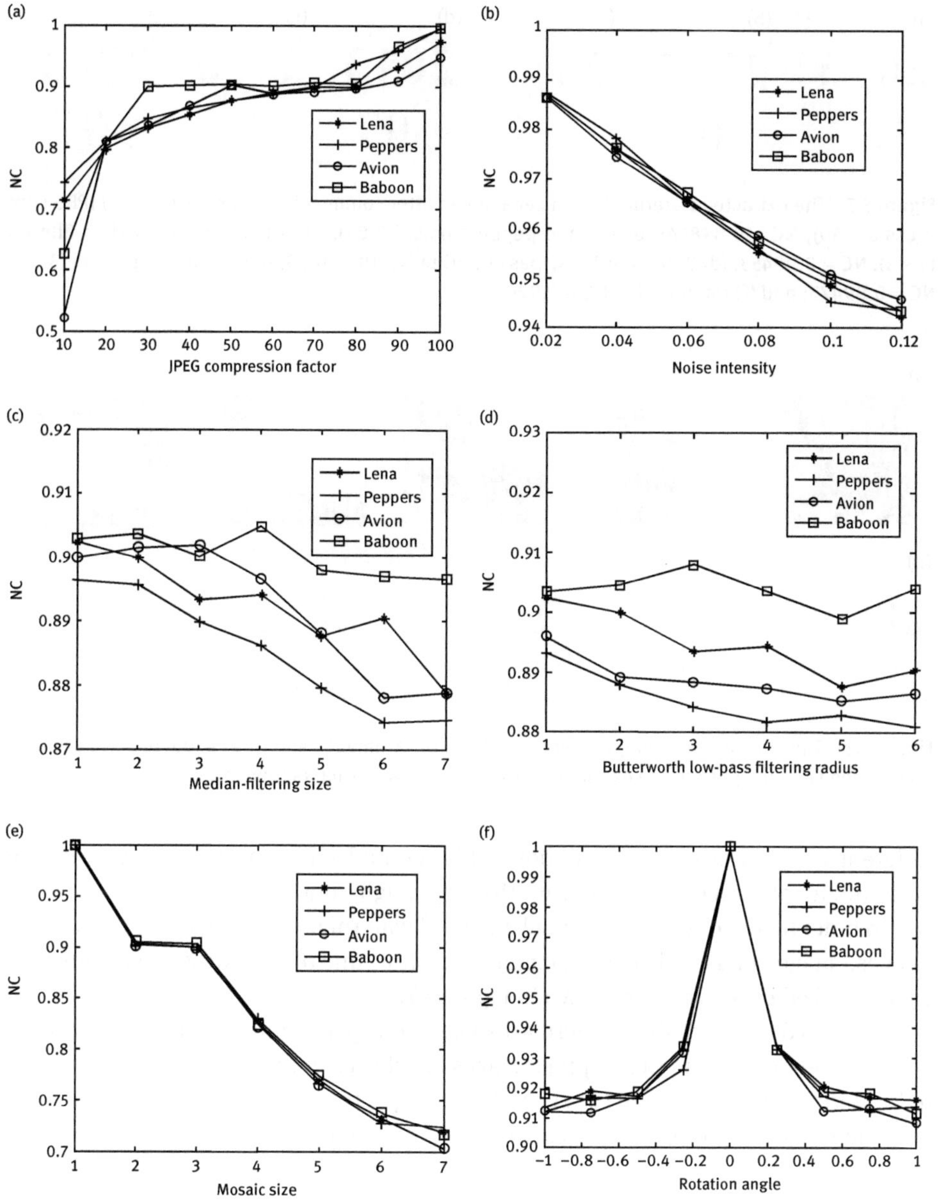

Figure 5.6: The extracted watermark after different attacks: (a) JPEG compression, (b) salt and peppers noise, (c) median filtering, (d) Butterworth low-pass filtering, (e) Mosaic piecing, and (f) rotation.

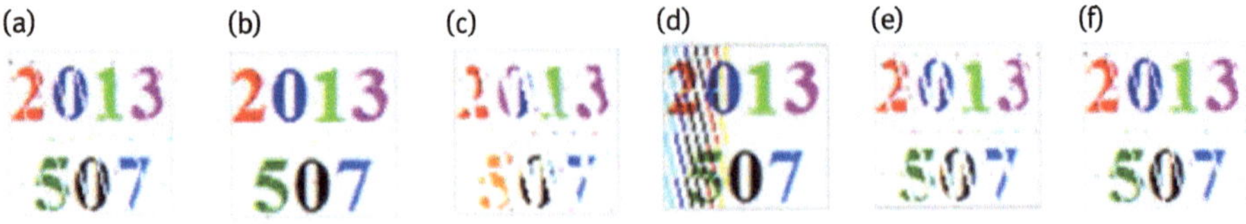

Figure 5.7: The extracted watermark from Lena image after common image processing: (a) JPEG compression (50), NC = 0.89854; (b) salt and peppers noise (0.02), NC = 0.98491; (c) median filtering (3 × 3), NC = 0.90453; (d) Butterworth low-pass filtering (4), NC = 0.89365; (e) mosaic piecing (3 × 3), NC = 0.89473; and (f) rotation (1°), NC = 0.92016.

Figure 5.8: The results of cropped watermarked Lena image and the extracted watermarks: (a) cropped Lena images and (b) extracted corresponding watermarks from (a).

different rotation angles, which shows this method has good robustness with the small angle rotation but not well with the large angle rotation.

In order to demonstrate the visual effects of watermark extracted from all attacked images, image Lena is taken as an example to show one of the extracting results from each attacked case, which is shown in Figure 5.7.

Figure 5.8 is the result of watermarked Lena images cropped in different positions with different sizes, and the cropped images are shown in the first row and the extracted watermarks are shown in the second row, respectively, which shows the proposed watermarking has strong robustness against the cropping attack.

5.4.3 Comparison with Some Related Color Watermarking Algorithms

In order to further test the robustness of the proposed watermarking method, image Lena is used as the host image to compare with the method in Ref. [130], and the results of NC are given in Table 5.1. We can see from Table 5.1 that this proposed method in this chapter has good robustness against many attacks, which also shows the validity of this proposed algorithm.

Table 5.1: The NC comparison between the proposed method and the method in Ref. [130] under many attacks.

Attack	Parameter	Method [130]	Proposed
Salt and pepper noise	0.02	0.56979	0.98712
	0.04	0.33502	0.97622
	0.06	0.27242	0.96526
	0.08	0.23636	0.95549
	0.1	0.21502	0.94867
Contrast adjustment	+2	0.89129	1.00000
JPEG compression	40	0.72571	0.85316
	50	0.77996	0.87643
	60	0.83923	0.89041
	70	0.90461	0.89929
	80	0.95254	0.89863
	90	0.97719	0.90046
Median filtering	1×1	0.99365	0.89999
	2×2	0.79198	0.90144
	3×3	0.71024	0.90180
	4×4	0.54596	0.89658
	5×5	0.50191	0.88812
Blurring	0.1	1.0000	1.00000
	1	0.27017	0.86106
Sharpening	0.1	0.85027	1.00000
	0.2	0.78075	0.96106

5.5 Conclusion

In this chapter, a new blind digital image watermarking algorithm based on IWT and state coding is proposed to embed color image watermark with more information. When embedding watermark, one-level IWT is performed on three components R, G, and B from the host image, then low-frequency coefficient is obtained and modified to improve the capacity according to watermark information and the proposed state coding; when extracting watermark, the watermark can be extracted by state coding, which can directly extract watermark without the host image and the original watermark. Experimental results show that the proposed scheme can embed 24-bit color image watermark of size 64×64 to 24-bit color host image of size 512×512, and the watermarked image has good invisibility of watermark, but the robustness of the watermark is not good since the right watermark cannot be guaranteed to extract when the image pixel is obviously changed under some attacks. Hence, the problem, which on how to improve the watermark performance and the watermark invisibility under the premise of meeting the requirements of the robustness, will be further researched in Chapter 6.

6 The Color Image Watermarking Algorithm Based on Singular Value Decomposition

In order to effectively improve the watermark invisibility when embedding color image watermark into color host image, a novel watermarking method based on the optimization compensation of singular value decomposition (SVD) is proposed in this chapter. First, the pixel block of size 4×4 is decomposed by SVD, and the watermark bit is embedded into the block by modifying the second row of the first column and the third row of the first column entry of matrix U. Then the embedded block is compensated by the optimization operation, which further improves the invisibility of embedded watermark. When the watermarked image is under attack, the embedded watermark can be extracted from the attacked images according to the relation between the modified entries of matrix U without restoring to the original data. Moreover, the proposed method overcomes the problem of false-positive detection and has strong robustness against common image processing.

6.1 Introduction

A blind watermarking method of dual-color images is proposed in Chapter 5, which can embed huge volume of color watermark information into the color host image with good invisibility of watermark at the expense of robustness as the premise. Hence, the method is very suitable for strong invisibility but weak robustness. Obviously, it is worthwhile to pay attention to how to improve the watermark invisibility under the premise of balancing the robustness of watermark.

As we know, more information can be embedded into the host image by the transform domain method, which, at the same time, has better robustness for common image processing operation. However, its computational complexity is much larger than that of the spatial method. The watermark information can be embedded into the host image in the spatial domain, but the spatial watermarking method has poor robustness against common image processing operations or attacks. In order to overcome the above drawbacks, the watermarking method based on SVD is becoming one of the hot spots in current research.

SVD, as a kind of strategy which is used to find the location of watermark embedding in the transform domain, has been earlier proposed by Liu et al. [131]. Then, many improved methods has been proposed, which can be roughly divided into three improved directions: (1) Combining some encryption methods or other embedding watermark method with SVD to complete the procedure of embedding watermark [132, 133], which is close to the original embedding method and just enhances the security of the method. (2) SVD and other transform domain methods are combined to obtain singular value with better robustness [134–136]. Relatively, the operating

DOI 10.1515/9783110487732-006

execution time of this combination method is longer than that of the former, and it is not easy to achieve the method with the hardware. (3) Because SVD was proposed to perform on the whole image at first, which was not satisfied whether on the safety or watermark capacity, it brought up to an idea that first the image is divided into nonoverlapping sub-blocks, then performing SVD on each sub-blocks [137–140]. This method greatly improves the performance of the original embedding method and gradually becomes a main direction for solving watermark problems by SVD [141–147].

Based on the research of the recent works, it is found that most of the methods based on SVD still exist problem of false-positive detection [148], and the cause of the problem is just embedding the singular value of the watermark W into the host image [149–152]. That is to say, when the singular value of the watermark W is $W = U_W D_W V_W^T$, only its singular value matrix D_W is embedded into the host image, while the orthogonal matrices U_W and V_W of the original watermark are not. In the extraction process, only singular value matrix D_W is extracted, but the U_W and V_W can be simply provided by the owner. However, the orthogonal matrices U_W and V_W contain the major information of the image [153], and any attackers can provide a pair of fake orthogonal matrices U_W and V_W and can claim that his watermark is embedded in the host image. In order to overcome this drawback, Chang et al. [154] proposed a block-based watermarking algorithm, in which the image was divided into several pixel blocks, and then the entries in matrix U of each block were modified to embed watermark. Although this method could reduce the number of the modified pixel from N^2 to $2N$, the modified quantity of each pixel is larger. Therefore, Fan et al. [155] further considered to modify the entries in the first column of matrix U (or matrix V) to embed the watermark and use the entries in the matrix V (or matrix U) to compensate the visual distortion, and the compensate method reduced the modifier and improved the robustness and invisibility of watermark; however, the number of the modified pixels is increased to N^2 to some extent since the existence of compensation operation, which will change the other pixels that should not have been modified. The detailed reasons will be explained in Section 6.3.

According to the above discussion, this chapter proposes an improved scheme to further optimize the compensation method based on SVD for embedding color watermark into the color host image. First, divide the original color image into nonoverlapping pixel block of size 4×4, perform SVD on it, and modify the entry in the first column of the second row and the one in the first column of the third row of the matrix U to embed the watermark; then, the matrix V is compensated by the proposed solutions and the final watermarked block which has high invisibility of watermark is obtained. The relation between the modified entries in matrix U can be well preserved, which in turn to extract the embedded watermark without the original image. Moreover, the proposed watermarking method in this chapter completely overcomes the problem of false-positive detection based on SVD method and improves the invisibility of the watermark by the proposed SVD based on optimization compensation.

6.2 The SVD of Image Block and the Compensation Optimization Method

6.2.1 The SVD of Image Block

From the perspective of linear algebra, a digital image can be regarded as a matrix composed of a lot of nonnegative scalar entries. $I \in R^{N\times N}$ is used to represent this image matrix, and R refers to a real number field so that I can be represented as

$$I = UDV^T = \begin{bmatrix} u_{1,1} & \cdots & u_{1,N} \\ u_{2,1} & \cdots & u_{2,N} \\ \vdots & \ddots & \vdots \\ u_{N,1} & \cdots & u_{N,N} \end{bmatrix} \begin{bmatrix} \lambda_1 & 0 & \cdots & 0 \\ 0 & \lambda_2 & \cdots & 0 \\ \vdots & \vdots & \ddots & 0 \\ 0 & 0 & \cdots & \lambda_N \end{bmatrix} \begin{bmatrix} v_{1,1} & \cdots & v_{1,N} \\ v_{2,1} & \cdots & v_{2,N} \\ \vdots & \ddots & \vdots \\ v_{N,1} & \cdots & v_{N,N} \end{bmatrix}, \tag{6.1}$$

where $U \in R^{N\times N}$ and $V \in R^{N\times N}$ are orthogonal matrices, $D \in R^{N\times N}$ is a matrix that all of the entries on off-diagonal are zero, and the entries on its diagonal meet:

$$\lambda_1 \geq \lambda_2 \geq \cdots \geq \lambda_r > \lambda_{r+1} = \cdots = \lambda_N = 0, \tag{6.2}$$

where r is the rank of I. It is equal to the number of the nonzero singular values; λ_i is referred to the singular value of I, which is a square root of II^T eigenvalue. Factorization UDV^T is called SVD of I. Because

$$II^T = UDV^T VD^T U^T = UDD^T U^T, \quad I^T I = VD^T U^T UDV^T = VD^T DV^T. \tag{6.3}$$

So the column vector of matrix U is the eigenvector of II^T, the column vector of matrix V is the eigenvector of $I^T I$, and the corresponding eigenvalues are the square of the singular values of I.

Since the research object is a digital image represented by matrix, in order to represent the meaning of the SVD more clearly, a detailed explanation of the eq. (6.1) is given as follows:

$$I = UDV^T = [U_1, U_2, \cdots, U_N] \begin{bmatrix} \lambda_1 & 0 & \cdots & 0 \\ 0 & \lambda_2 & \cdots & 0 \\ \vdots & \vdots & \ddots & 0 \\ 0 & 0 & \cdots & \lambda_N \end{bmatrix} [V_1, V_2, \cdots, V_N]^T, \tag{6.4}$$

where $[U_1, U_2, \cdots, U_N]$ and $[V_1, V_2, \cdots, V_N]$ denote its left and right eigenvectors, respectively.

According to eq. (6.4), we can get the spectral resolution of the original image I:

$$I = U_1\lambda_1 V_1^T + U_2\lambda_2 V_2^T + \cdots + U_N\lambda_N V_N^T = \lambda_1 U_1 V_1^T + \lambda_2 U_2 V_2^T + \cdots + \lambda_N U_N V_N^T. \quad (6.5)$$

It can be seen from eq. (6.5) that after decomposing the singular value, the original image I can be represented to the form of superposition of the sum of N sub-images such as $\lambda_1 U_1 V_1^T$, $\lambda_2 U_2 V_2^T, \ldots, \lambda_N U_N V_N^T$. These sub-images are layers of frame images of original image, and the singular value can be seen as weighted value when these frame images reconstruct the original image. It can be seen from this decomposition that matrices U, V store the geometry information of the image, while the singular values store the brightness information of the image.

According to the definitions and decomposition of singular value, it can be seen that the singular value has the following properties [156]:

1. The representativeness of eigenvector of SVD
   It can be seen from the above analysis that the original image has corresponding relations with its eigenvectors, so the eigenvector of SVD can be used to describe a two-dimensional image. As the image's gray-level information changes in a certain range, the eigenvectors will not have big changes. Hence, the eigenvector is not sensitive to the changes of gray scale caused by image noise and different image illumination conditions, which has certain stabilities. Thus, it can reduce the requirements of the image preprocessing, and the eigenvector of SVD has the stability representative toward the original image.
2. The transposition invariance of eigenvector of SVD
   From the definition and formula of SVD, it's easy to see that the eigenvector of SVD will not change when performing transpose operations on the image.
3. The rotation invariance of eigenvector of SVD
   The eigenvector of singular value will not change when performing twiddle operation on the image.
4. The shifting invariance of eigenvector of SVD
   The movement of the image, that is performing the permutation operation on the row or column of original image matrix, cannot change the eigenvector of SVD.
5. The main information stored in several ahead singular values.
   In the singular value sequences obtained by SVD, the ahead singular values are much larger than others. Thus, it can restore the image in the situation even when ignoring other singular values.

Supposing a 4×4 matrix A is the block matrix of the host image, its SVD can be represented by

$$A = \begin{bmatrix} A_1 & A_2 & A_3 & A_4 \\ A_5 & A_6 & A_7 & A_8 \\ A_9 & A_{10} & A_{11} & A_{12} \\ A_{13} & A_{14} & A_{15} & A_{16} \end{bmatrix} = UDV^T$$

$$= [U_1, U_2, U_3, U_4] \begin{bmatrix} \lambda_1 & 0 & 0 & 0 \\ 0 & \lambda_2 & 0 & 0 \\ 0 & 0 & \lambda_3 & 0 \\ 0 & 0 & 0 & \lambda_4 \end{bmatrix} [V_1, V_2, V_3, V_4]^T$$

$$= \begin{bmatrix} u_1 & u_2 & u_3 & u_4 \\ u_5 & u_6 & u_7 & u_8 \\ u_9 & u_{10} & u_{11} & u_{12} \\ u_{13} & u_{14} & u_{15} & u_{16} \end{bmatrix} \begin{bmatrix} \lambda_1 & 0 & 0 & 0 \\ 0 & \lambda_2 & 0 & 0 \\ 0 & 0 & \lambda_3 & 0 \\ 0 & 0 & 0 & \lambda_4 \end{bmatrix} \begin{bmatrix} v_1 & v_2 & v_3 & v_4 \\ v_5 & v_6 & v_7 & v_8 \\ v_9 & v_{10} & v_{11} & v_{12} \\ v_{13} & v_{14} & v_{15} & v_{16} \end{bmatrix}^T . \tag{6.6}$$

Performing matrix multiplication on U, D, V^T, each pixel can be calculated as follows:

$$A_1 = u_1\lambda_1v_1 + u_2\lambda_2v_2 + u_3\lambda_3v_3 + u_4\lambda_4v_4, \tag{6.7}$$
$$A_2 = u_1\lambda_1v_5 + u_2\lambda_2v_6 + u_3\lambda_3v_7 + u_4\lambda_4v_8, \tag{6.8}$$
$$A_3 = u_1\lambda_1v_9 + u_2\lambda_2v_{10} + u_3\lambda_3v_{11} + u_4\lambda_4v_{12}, \tag{6.9}$$
$$A_4 = u_1\lambda_1v_{13} + u_2\lambda_2v_{14} + u_3\lambda_3v_{15} + u_4\lambda_4v_{16}, \tag{6.10}$$
$$A_5 = u_5\lambda_1v_1 + u_6\lambda_2v_2 + u_7\lambda_3v_3 + u_8\lambda_4v_4, \tag{6.11}$$
$$A_6 = u_5\lambda_1v_5 + u_6\lambda_2v_6 + u_7\lambda_3v_7 + u_8\lambda_4v_8, \tag{6.12}$$
$$A_7 = u_5\lambda_1v_9 + u_6\lambda_2v_{10} + u_7\lambda_3v_{11} + u_8\lambda_4v_{12}, \tag{6.13}$$
$$A_8 = u_5\lambda_1v_{13} + u_6\lambda_2v_{14} + u_7\lambda_3v_{15} + u_8\lambda_4v_{16}, \tag{6.14}$$
$$A_9 = u_9\lambda_1v_1 + u_{10}\lambda_2v_2 + u_{11}\lambda_3v_3 + u_{12}\lambda_4v_4, \tag{6.15}$$
$$A_{10} = u_9\lambda_1v_5 + u_{10}\lambda_2v_6 + u_{11}\lambda_3v_7 + u_{12}\lambda_4v_8, \tag{6.16}$$
$$A_{11} = u_9\lambda_1v_9 + u_{10}\lambda_2v_{10} + u_{11}\lambda_3v_{11} + u_{12}\lambda_4v_{12}, \tag{6.17}$$
$$A_{12} = u_9\lambda_1v_{13} + u_{10}\lambda_2v_{14} + u_{11}\lambda_3v_{15} + u_{12}\lambda_4v_{16}, \tag{6.18}$$
$$A_{13} = u_{13}\lambda_1v_1 + u_{14}\lambda_2v_2 + u_{15}\lambda_3c_3 + a_{16}\lambda_4c_4, \tag{6.19}$$
$$A_{14} = u_{13}\lambda_1v_5 + u_{14}\lambda_2v_6 + u_{15}\lambda_3v_7 + u_{16}\lambda_4v_8, \tag{6.20}$$
$$A_{15} = u_{13}\lambda_1v_9 + u_{14}\lambda_2v_{10} + u_{15}\lambda_3v_{11} + u_{16}\lambda_4v_{12}, \tag{6.21}$$
$$A_{16} = u_{13}\lambda_1v_{13} + u_{14}\lambda_2v_{14} + u_{15}\lambda_3v_{15} + u_{16}\lambda_4v_{16}. \tag{6.22}$$

It can be seen from the above results that all pixels A_i will be changed when any eigenvalue λ_i changed, and this change will cause greater change of pixel when there are multiple eigenvalues modified at the same time, which will directly affect the invisibility of the watermark.

In Ref. [130], the modifying eigenvalue method was proposed to embed watermark. For example, assume that eigenvalues from λ_1 to λ_{16} of a 16×16 pixel block are 3,165.613, 457.5041, 31.54169, 9.85382997, 5.796001, 4.991171, 3.688464, 2.544742, 2.064232, 1.691997, 1.130058, 1.074023, 0.819865, 0.448544, 0.37897, and 0.101045. According to the method in Ref. [130], the eigenvalues will accordingly change into 3,165.613, 457.5041, 31.54169, 9.85382997, 0, 0, 0, 0, 0, 0, 0, 0, 0, 0, 0, when the embedded watermark information is 0; thus, 12 eigenvalues have been changed. The quality of watermark image will be seriously affected by the big changes of all the pixels of host image, so the disadvantage of the method is obvious.

It can be found from further research that the matrix U has the following features after performing SVD: All the entries of the first column have the same sign and their values are very close. For example, get a 4×4 pixel block matrix A from a digital image (the pixels of the digital image have nonnegativity and the neighbor entries are very similar), as follows:

$$A = \begin{bmatrix} 201 & 201 & 199 & 198 \\ 202 & 202 & 199 & 199 \\ 203 & 203 & 201 & 201 \\ 203 & 204 & 202 & 202 \end{bmatrix}. \qquad (6.23)$$

Perform SVD on matrix A, and the orthogonal matrix U is obtained as follows:

$$U = \begin{bmatrix} -0.49627 & -0.32583 & 0.80128 & 0.07426 \\ -0.49813 & -0.57109 & -0.57014 & 0.31727 \\ -0.50185 & 0.14742 & -0.17355 & -0.83444 \\ -0.50371 & 0.73890 & -0.05270 & 0.44445 \end{bmatrix}. \qquad (6.24)$$

As can be seen from matrix U, the first column entries, that is, u_1, u_5, u_9, u_{13}, are of the same numerical sign and the differences between them are very small. Suppose one matrix is composed of entry u_m in the first column of each matrix U and another one is composed of another entry u_n in the first column of each matrix U, where $m \neq n$. Then the similarities between u_m and u_n are calculated by normalized cross-correlation (NC). Table 6.1 shows the results of using many standard images in the experiment. It can be seen from the table that the average value of NC(u_5, u_9) is 0.9886, which means u_5 and u_9 are the most similar entries in the first column of the 4×4 matrix U for many standard images. This feature can be explored for embedding digital watermark in Section 6.3.

6.2.2 The Proposed Optimization Compensation Method of SVD

As shown in eqs (6.11–6.18), modifying the values of u_5 and u_9 will change the values of $A_i (i = 5, 6, \ldots, 12)$, which decreases the visual effect of the embedded watermark [157].

Table 6.1: The similarity between different elements in first column of U matrix after SVD (NC).

Image	NC(u_1,u_5)	NC(u_1,u_9)	NC(u_1,u_{13})	NC(u_5,u_9)	NC(u_5,u_{13})	NC(u_9,u_{13})
Lena	0.9934	0.9886	0.9871	0.9969	0.9901	0.9940
House	0.9966	0.9942	0.9935	0.9990	0.9949	0.9969
Peppers	0.9673	0.9482	0.9444	0.9871	0.9554	0.9692
Avion	0.9921	0.9873	0.9815	0.9972	0.9884	0.9940
Baboon	0.9709	0.9589	0.9525	0.9796	0.9579	0.9716
Bear	0.9153	0.8848	0.8839	0.9564	0.9069	0.9341
Kid	0.9942	0.9896	0.9823	0.9962	0.9852	0.9919
Sailboat	0.9879	0.9798	0.9779	0.9967	0.9796	0.9876
Barbara	0.9882	0.9785	0.9728	0.9947	0.9814	0.9913
Couple	0.9323	0.9006	0.8907	0.9818	0.9219	0.9538
Average	0.9738	0.9610	0.9567	0.9886	0.9662	0.9784

Therefore, a compensation method was proposed in Ref. [155], which adopted matrix V (or matrix U) to compensate the visible distortion when embedding watermark into matrix U (or matrix V) to some extent. That was an effective compensation method to enhance the watermarking performance to some extent. However, we find that the proposed compensation method [155] can be further optimized. The reason can be explained by the following examples.

First, the results of modified u_5, u_9 have caused abrupt distortion not only on the pixels of $A_i(i = 5, 6, \ldots, 12)$, but also on the ones of other entries by compensation operation. Therefore, the invisibility of the embedded watermark will be further decreased. In Figure 6.1, we define the original matrix as A, the matrix with modified u_5, u_9 as A^*, the compensated matrix as A^{**}, and the final embedded watermark matrix as A^{***}. As can be seen from the matrices A and A^{***} of matrix 1 in Figure 6.1, the values of A_8 and A_{12} of matrix 1 have been changed from 0 to 3, 8 to 4, respectively, when embedding watermark bit is "0." However, A_{16} is also changed from 23 to 21 by the compensation method in Ref. [155], which further distorts the whole visual quality of the watermarked image. This is not what we want to see.

Second, the total modified values of $A_i(i = 5, 6, \ldots, 12)$ should be optimized to minimum value, which can enhance the visibility of embedded watermark. This chapter can explain this by the variation of A_2 in matrix 2 of Figure 6.1; its value is changed from 99 to 100 after being compensated by the method in Ref. [155], but the compensated value 100 is not to be used in its final matrix A^{***}. Obviously, this is another drawback of the method in Ref. [155]. Therefore, it is necessary to improve and optimize the method in Ref. [155].

The detailed initial compensation method for matrix V can be referred to the method in Ref. [155]. In order to optimize the result of compensation, this chapter

	Matrix 1		Matrix 2	
	Fan et al. [155]	Proposed	Fan et al. [155]	Proposed
A	$\begin{bmatrix} 0 & 0 & 0 & 1 \\ 0 & 0 & 0 & \underline{0} \\ 0 & 0 & 1 & \underline{8} \\ 0 & 0 & 3 & \underline{23} \end{bmatrix}$	$\begin{bmatrix} 0 & 0 & 0 & 1 \\ 0 & 0 & 0 & 0 \\ 0 & 0 & 1 & 8 \\ 0 & 0 & 3 & 23 \end{bmatrix}$	$\begin{bmatrix} 112 & 91 & 186 & 198 \\ \underline{108} & 62 & 147 & 217 \\ 111 & 64 & 135 & 194 \\ 118 & 119 & 92 & 174 \end{bmatrix}$	$\begin{bmatrix} 112 & 91 & 186 & 198 \\ 108 & 62 & 147 & 217 \\ 111 & 64 & 135 & 194 \\ 118 & 119 & 92 & 174 \end{bmatrix}$
A^*	$\begin{bmatrix} 0 & 0 & 0 & 1 \\ 0 & 0 & 0 & \underline{4} \\ 0 & 0 & 1 & \underline{4} \\ 0 & 0 & 3 & \underline{23} \end{bmatrix}$	$\begin{bmatrix} 0 & 0 & 0 & 1 \\ 0 & 0 & 0 & 4 \\ 0 & 0 & 1 & 4 \\ 0 & 0 & 3 & 23 \end{bmatrix}$	$\begin{bmatrix} 112 & 91 & 186 & 198 \\ \underline{99} & 56 & 136 & 202 \\ 120 & 70 & 146 & 209 \\ 118 & 119 & 92 & 174 \end{bmatrix}$	$\begin{bmatrix} 112 & 91 & 186 & 198 \\ 99 & 56 & 136 & 202 \\ 120 & 70 & 146 & 209 \\ 118 & 119 & 92 & 174 \end{bmatrix}$
A^{**}	$\begin{bmatrix} 0 & 0 & 0 & 1 \\ 0 & 0 & 0 & \underline{3} \\ 0 & 0 & 0 & \underline{4} \\ 0 & 0 & 3 & \underline{21} \end{bmatrix}$	$\begin{bmatrix} 0 & 0 & 0 & 1 \\ 0 & 0 & 0 & 3 \\ 0 & 0 & 0 & 4 \\ 0 & 0 & 3 & 21 \end{bmatrix}$	$\begin{bmatrix} 112 & 91 & 186 & 199 \\ \underline{100} & 56 & 137 & 203 \\ 120 & 71 & 146 & 210 \\ 118 & 119 & 92 & 175 \end{bmatrix}$	$\begin{bmatrix} 112 & 91 & 186 & 199 \\ 100 & 56 & 137 & 203 \\ 120 & 71 & 146 & 210 \\ 118 & 119 & 92 & 175 \end{bmatrix}$
A^{***}	$\begin{bmatrix} 0 & 0 & 0 & 1 \\ 0 & 0 & 0 & \underline{3} \\ 0 & 0 & 0 & \underline{4} \\ 0 & 0 & 3 & \underline{21} \end{bmatrix}$	$\begin{bmatrix} 0 & 0 & 0 & 1 \\ 0 & 0 & 0 & 3 \\ 0 & 0 & 0 & 4 \\ 0 & 0 & 3 & \underline{23} \end{bmatrix}$	$\begin{bmatrix} 112 & 91 & 186 & 198 \\ \underline{99} & 56 & 136 & 202 \\ 120 & 70 & 146 & 209 \\ 118 & 119 & 92 & 174 \end{bmatrix}$	$\begin{bmatrix} 112 & 91 & 186 & 198 \\ \underline{100} & 56 & 137 & 203 \\ 120 & 70 & 146 & 209 \\ 118 & 119 & 92 & 174 \end{bmatrix}$
CM	30	26	926	859

Figure 6.1: The comparison result of modified energy between different methods.

uses eq. (6.25) to further optimize the compensation method [155] and obtains the final embedded watermark matrix A^{***}:

$$A_i^{***} = \begin{cases} A_i + \arg\min(|A_i^* - A_i|, |A_i^{**} - A_i|), & \text{if } 5 \le i \le 12 \\ A_i, & \text{otherwise} \end{cases}. \tag{6.25}$$

Figure 6.1 also gives three different states of A^*, A^{**}, and A^{***} of matrices 1 and 2 when watermark “0” is embedded by different methods and lists comparison results between different methods in terms of the changed magnitude (CM), which can be defined by eq. (6.26):

$$\text{CM} = \sum_{i=1}^{16} (A_i^{***} - A_i)^2. \tag{6.26}$$

The lower the CM is, the better invisibility of the watermarking method has. As shown in Figure 6.1, the CM values obtained by the proposed method are fewer than those obtained by the method in Ref. [155]. Therefore, the proposed method is superior to the method in Ref. [155].

6.3 The Proposed Color Image Watermarking Algorithm

In this section, we will explore the property of matrix U mentioned in Section 6.2 to embed the color image watermark and use the proposed compensation method to optimize the embedded watermark.

6.3.1 Watermark Embedding

The procedure of embedding color image digital watermark is shown in Figure 6.2, and the detailed steps are introduced as follows:

1. Preprocessing the color watermark
   The three-dimensional original color watermark image W is first divided into R, G, and B components by dimension-reduction operation to form two-dimensional component watermarks W_i $(i = 1, 2, 3)$, which represents R, G, and B component watermarks, respectively. In order to enhance the security and robustness of the watermark, each component watermark is permuted by Arnold transform with private key KA_i $(i = 1, 2, 3)$ and converted to 8-bit binary sequence [158].
2. The block processing of the host image
   The host image H is also divided into three components H_i $(i = 1, 2, 3)$, which represent the R, G, and B components, respectively; At the same time, each component image is divided into nonoverlapping blocks of size 4×4.
3. Selecting the embedding block
   Use the pseudorandom sequence based on the key K_i $(i = 1, 2, 3)$ to select the embedding block in component image H_i to embed watermark components W_i.
4. SVD transformation

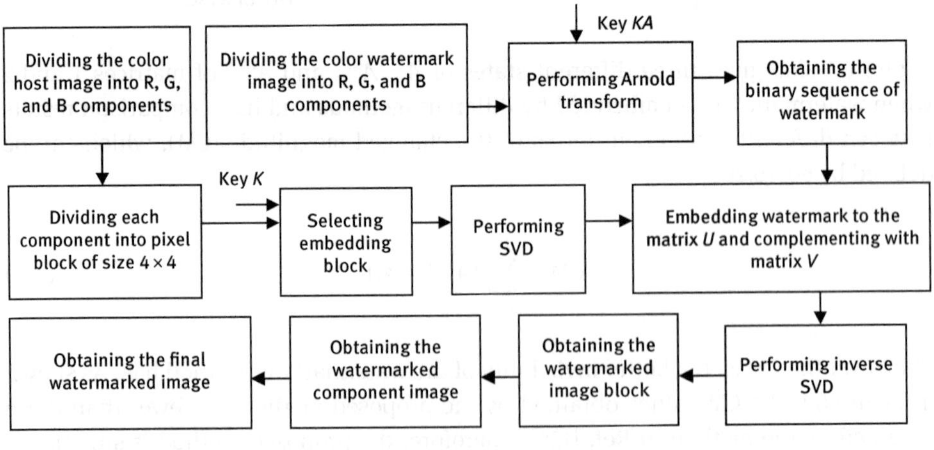

Figure 6.2: The diagram of watermark embedding.

In order to obtain the matrix U, each selected block is transformed by SVD according to eq. (6.6).

5. Embedding watermark
The watermark is embedded by changing the relation between the entry u_5, which is in the first column of the second row, and the entry u_9 which is in the first column of the third row. If the embedded binary information is "1," the value of $u_5 - u_9$ should be negative and its magnitude is greater than a certain threshold value T. If the embedded binary information is "0," the value of $u_5 - u_9$ should be positive and its magnitude is also greater than a certain threshold value T. When these two conditions are violated, the values of u_5 and u_9 should be, respectively, changed to u_5^* and u_9^* based on the following rules in eqs (6.27) and (6.28):

$$\text{if } w = 1 \quad \text{and} \quad |u_5 - u_9| < T, \qquad \begin{cases} u_5^* = \text{sign}(u_5) \times (U_{\text{avg}} + T/2) \\ u_9^* = \text{sign}(u_9) \times (U_{\text{avg}} - T/2) \end{cases}, \tag{6.27}$$

$$\text{if } w = 0 \quad \text{and} \quad |u_5 - u_9| < T, \qquad \begin{cases} u_5^* = \text{sign}(u_5) \times (U_{\text{avg}} - T/2) \\ u_9^* = \text{sign}(u_9) \times (U_{\text{avg}} + T/2) \end{cases}, \tag{6.28}$$

where w represents the embedded binary watermarking information, $|x|$ refers to the absolute value of x, sign(x) stands for the numerical sign of x, and $U_{\text{avg}} = (|u_5| + |u_9|)/2$.

6. Performing the proposed compensation optimization method
The embedded pixel block is compensated and optimized by the compensated optimization method proposed in Section 6.5.

7. Inverse SVD transformation
Use eq. (6.29) to perform inverse SVD transform on the modified and compensated matrices U^* and V^* , and obtain the watermarked pixel block I^*:

$$I^* = U^* D V^{*^T}. \tag{6.29}$$

8. Looping
Repeat steps 4–7 until all watermark information is embedded into the selected pixel block. Finally, recombining the watermarked component images R, G, and B to get watermarked image H^*.

6.3.2 Watermark Extraction

The procedure of watermark extraction is shown in Figure 6.3, and the detailed steps are explained as follows:

1. Preprocessing the watermarked image.
The watermarked image H^* is divided into R, G, and B three-component images, which are further divided into watermarked image blocks with size of 4×4.

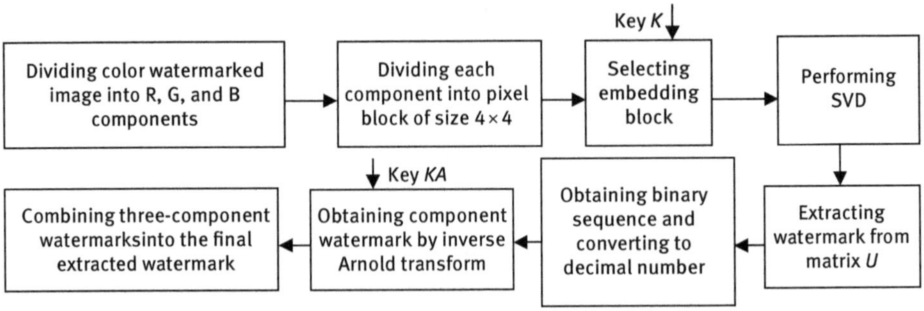

Figure 6.3: The diagram of watermark extraction.

2. Selecting the watermarked image block
   Use pseudorandom sequences based on the key K_i ($i = 1, 2, 3$) to select the embedded block of component image H_i^* for extracting the embedded watermark component W_i^*
3. SVD transformation
   Perform SVD transformation on the image block to obtain the orthogonal matrix U^*.
4. Extracting watermark
   Use the relationship between the entries u_5^* and u_9^* in the first column of the matrix U^* to extract the watermark according to eq. (6.30):

$$w^* = \begin{cases} 0, & \text{if } u_5^* > u_9^* \\ 1, & \text{if } u_5^* \le u_9^* \end{cases} . \tag{6.30}$$

5. Looping
   Repeat the above steps 2–4 until all the watermarked pixel blocks are processed. Then these extracted bit values are partitioned into 8-bit groups and converted to decimal number.
6. Recombining
   Use the inverse Arnold transform based on key KA_i ($i = 1, 2, 3$) to transform each component watermark, and recombine them to form the final extracted watermark W^*.

It is noted that the proposed method of extracting watermark does not need the original watermark image and the original host image, which belongs to blind watermarking extraction technology.

6.4 Experimental Results and Analysis

In the following experiment, two 24-bit true color images (Baboon and Avion) with size of 512 × 51, as shown in Figure 6.4(a) and 6.4(b), are selected as the original host

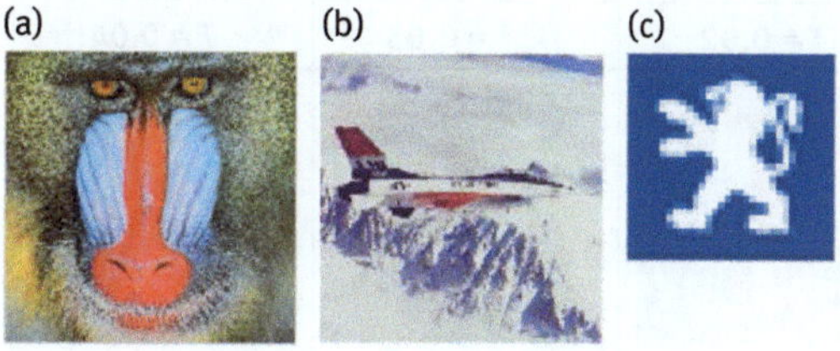

Figure 6.4: Original host image: (a) Baboon and (b) Avion; original watermark image: (c) Peugeot logo.

image. In addition, one 24-bit true color image of size 32 × 32, as shown in Figure 6.4(c), is used as an original watermark.

In this chapter, structural similarity index measurement (SSIM) is used to evaluate the similarity between the original color host image I and the watermarked color image I^*; in other words, it is used to evaluate the invisibility of the watermark; at the same time, NC as a kind of objective measuring standard is used to evaluate the extracted watermark W^* and the original watermark W, which also can evaluate the robustness of the watermark.

6.4.1 Test of the Watermark Invisibility

To evaluate the invisibility of watermark, embed the color image watermark shown in Figure 6.4(c) into the host images shown in Figure 6.4(a) and 6.4(b), and compare the SSIM in different threshold values with the method described in Ref. [155]. Figure 6.5 shows the respective SSIM values, and the results show that the proposed compensation optimization method with different threshold values has good watermark invisibility, which meets the expected purpose.

In addition, Figure 6.6 shows the extracted watermarks and its NC values via using different methods without any attacks. It can be seen from the results that the extracted watermark has very good similarity with the increase of the threshold value T; thus, $T = 0.04$ is the best selection in the following experiments.

In the following experiments, various attacks, such as Joint Photographic Expert Group (JPEG) compression, adding noise, filtering, sharpening, scaling, blurring, rotation, cropping, and so on, are performed on the watermarked image to test the robustness of the proposed method, and compare with the methods in Refs [155] and [130].

6.4.2 Test of the Watermark Robustness

Nowadays, JPEG compression is a popular compression format, and JPEG compression attack is also a kind of effective attack method to verify the performance of watermarking method. In this experiment, the watermarked image is compressed with different compression factors ranging from 10 to 100 increasing in step of 10, and then

Image	Method	$T = 0.012$	$T = 0.02$	$T = 0.03$	$T = 0.04$
Baboon	Fan et al. [155]	 0.98766	 0.98513	 0.98109	 0.97631
	Proposed	 0.98788	 0.98534	 0.98130	 0.97646
Avion	Fan et al. [155]	 0.99108	 0.97834	 0.95655	 0.93231
	Proposed	 0.99113	 0.97928	 0.95838	 0.93304

Figure 6.5: The watermarked images obtained by different methods and their SSIM values.

the embedded watermark information is extracted from the compressed image. The bigger the compression factor is, the better the image quality after being compressed is, and the more easily to extract the embedded watermark. Figure 6.7 shows the experimental results when the compression factor is 30 and 90, respectively. Compared with other methods, the method proposed in this chapter has better robustness against JPEG compression.

Using the salt and peppers noise intensity is 2 % and 10 %, respectively, to attack the watermarked image, Figure 6.8 shows the NC value and visual effect of the extracted watermark. In addition, using the Gaussian noise whose mean value is 0, variances are 0.1 and 0.3, respectively, to attack the watermark image, Figure 6.9 shows the NC value and extracted watermark after Gaussian noising attack. It is shown from Figures 6.8 and 6.9 that the proposed watermarking method has better robustness against adding noise attack than other methods. Among them, the robustness of method [130] is weaker than the others; this is because the effects of noise attacks are larger on image pixel values and can directly affect the singular values, which is more

Image	Method	$T = 0.012$	$T = 0.02$	$T = 0.03$	$T = 0.04$
Baboon	Fan et al. [155]	 0.99975	 0.99993	 0.99993	 0.99993
	Proposed	 0.99901	 1.00000	 1.00000	 1.00000
Avion	Fan et al. [155]	 0.99957	 0.99997	 1.00000	 1.00000
	Proposed	 0.99939	 0.99938	 1.00000	 1.00000

Figure 6.6: The extracted watermarks and NC values via using different methods without any attacks.

Attack	Image	Fan et al. [155]	Golea et al. [130]	Proposed
JPEG(30)	Baboon	 0.79280	 0.65311	 0.81049
	Avion	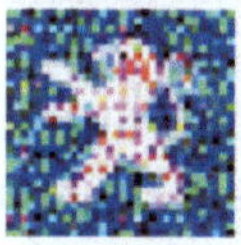 0.83975	 0.82043	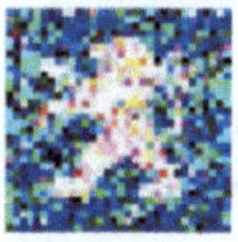 0.84941
JPEG(90)	Baboon	 0.99587	 0.92492	 0.99611
	Avion	 0.99732	 0.97595	 0.99886

Figure 6.7: The extracted watermarks and NC values via using different methods after JPEG compression attack.

Attack	Image	Fan et al. [155]	Golea et al. [130]	Proposed
Salt & Peppers noise (2%)	Baboon	 0.94802	 0.56979	 0.95366
	Avion	 0.95572	 0.52759	 0.97910
Salt & Peppers noise (10%)	Baboon	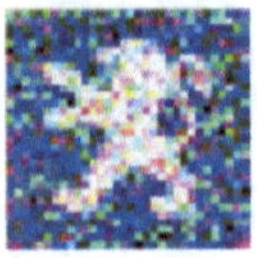 0.89004	0.27242	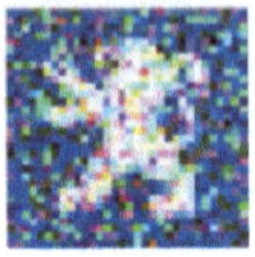 0.89027
	Avion	 0.90616	 0.21075	 0.92599

Figure 6.8: The extracted watermarks and NC values via using different methods after salt and peppers noising attack.

noticeable for the watermark information that is extracted by relying on the singular value.

Figure 6.10 shows the results of the median filtering attacks. From Figure 6.10, we can see that although the methods that list in the figure all show the weaker robustness, the proposed method is relatively better than others. The reason why the method lists in the figure represents weaker robustness is mainly because the method is based on partitioned image block and the partition size is not the same with the filter size.

In addition, the watermarked image is attacked by Butterworth low-pass filter whose cutoff frequency is 100 and filter order numbers are 1 and 3, respectively. Figure 6.11 shows the NC values of extracted watermarks and their visual effect. As can be seen from it, the extracted watermark by the proposed method has good robustness. With the increasing of the filter's orders, the damped speed of the resistance band amplitude expedites, which has larger influence to the watermarked image and more difficulties to extract the watermark.

Figure 6.12 shows the results of sharpening attack. In the procedure of image sharpening, the template operation is usually used, and the pixel values in the image that have big differences with neighbor pixels become more prominent after

Attack	Image	Fan et al. [155]	Golea et al. [130]	Proposed
Gaussian noise (0.1)	Baboon	 0.96323	 0.86001	 0.96726
	Avion	 0.98270	 0.81877	 0.98671
Gaussian noise (0.3)	Baboon	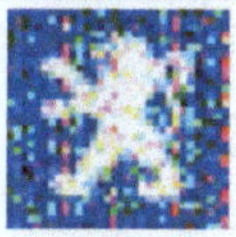 0.84644	 0.74694	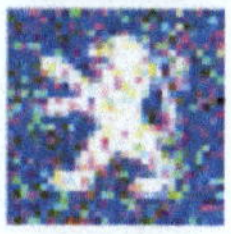 0.84828
	Avion	 0.51026	 0.65114	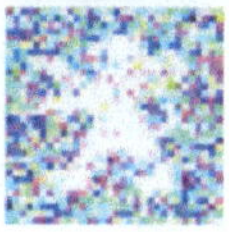 0.55691

Figure 6.9: The extracted watermarks and NC values via using different methods after Gaussian noising attack.

Attack	Image	Fan et al. [155]	Golea et al. [130]	Proposed
Median Filter (2×2)	Baboon	 0.69310	 0.65192	 0.72289
	Avion	 0.87758	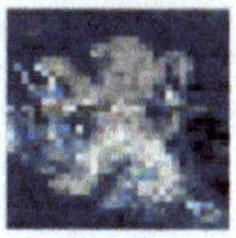 0.67150	 0.88433
Median Filter (3×3)	Baboon	 0.53841	 0.50744	 0.54619
	Avion	 0.50438	 0.53405	 0.54399

Figure 6.10: The extracted watermarks and NC values via using different methods after median filtering attack.

Attack	Image	Fan et al. [155]	Golea et al. [130]	Proposed
Low-pass Filter (100,1)	Baboon	 0.86449	 0.54766	 0.89987
	Avion	 0.92682	 0.58553	 0.95879
Low-pass Filter (100,3)	Baboon	 0.67236	 0.38175	 0.68958
	Avion	 0.83946	 0.46500	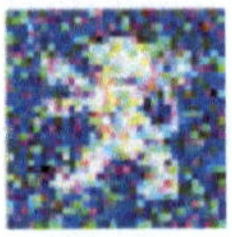 0.85599

Figure 6.11: The extracted watermarks and NC values via using different methods after low-pass filtering attack.

Attack	Image	Fan et al. [155]	Golea et al. [130]	Proposed
Sharpening (0.2)	Baboon	 0.99993	 0.84806	 0.99994
	Avion	 1.00000	 0.85027	 1.00000
Sharpening (1.0)	Baboon	 0.99904	 0.78075	 0.99916
	Avion	 0.99794	 0.62558	 0.99982

Figure 6.12: The extracted watermarks and NC values via using different methods after sharpening attack.

Attack	Image	Fan et al. [155]	Golea et al. [130]	Proposed
Scaling (400%)	Baboon	 0.93696	 0.83852	 0.94319
	Avion	 0.96868	 0.86885	 0.98137
Scaling (25%)	Baboon	 0.68716	 0.56979	 0.71741
	Avion	 0.85431	 0.61460	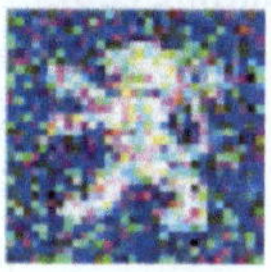 0.86793

Figure 6.13: The extracted watermarks and NC values via using different methods after scaling attack.

sharpening with the Laplace template. This chapter uses the USM sharpening in Photoshop with the sharpening radii of 0.2 and 1.0, respectively. It can be seen from Figure 6.12 that all of the other methods have very strong robustness except for Golea et al. [130].

Two different scaling operation images are performed on the watermarked during experiment, namely zoomed out by 400 % and zoomed in by 50 %. Figure 6.13 shows the experimental results. As shown in the figure, the method has very good robustness when it is zoomed out, while it generally has bad robustness when it is zoomed in; this is because when the image is under the magnification attack, the row (column) of the image will increase evenly, and the row (column) in the difference feature points may also increase accordingly. These difference feature points can guarantee the only one base point to extract watermark when the image is zoomed out, which makes the watermarking method has a strong robustness. When the image is zoomed in, the reduction of the image can make regular loss of the row (column) of the image, while the row (column) of difference feature points also may be lost. When the scaling proportion is less than 0.5, it does not always detect the base point of the difference feature

Attack	Image	Fan et al. [155]	Golea et al. [130]	Proposed
Blurring (0.2)	Baboon	 0.99987	 1.00000	 1.00000
	Avion	 0.99984	 0.57185	 1.00000
Blurring (1.0)	Baboon	 0.67205	 0.27017	 0.69519
	Avion	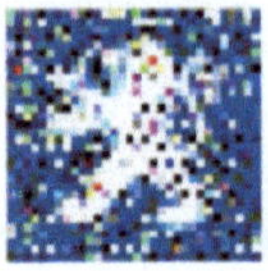 0.80154	 0.17850	 0.84759

Figure 6.14: The extracted watermarks and NC values via using different methods after blurring attack.

points, which generally affects the quality of the extracted watermark, which is also a problem that is worth to pay attention [159].

Two different blurring attacks are performed on the watermarked image with the fuzzy radii 0.2 and 1.0, respectively. Figure 6.14 shows the visual effect of the extracted watermark and the NC value. The larger the fuzzy radius is, the worse the robustness is; however, the proposed watermarking method has better robustness than other watermarking methods.

At the same time, two different rotation attacks are performed on the watermarked image. One is rotating the watermark image clockwise 5°, while the other is 30°, which contains the attacks such as rotation, scaling, cropping, and so on. For the rotated image, this chapter will rotate the watermarked image which is counterclockwise back to its original position and cut out an effective size. Figure 6.15 shows the results of the extracted watermark. In general, the relevant color watermarking method against rotating attack does not have strong robustness, especially large-angle rotation attacks. The method proposed in this chapter is slightly better than other methods.

Two cropping attacks with cropping proportions 25 % and 50 % are, respectively, used to crop the watermarked image. As the watermark in the method in Ref. [130]

Attack	Image	Fan et al. [155]	Golea et al. [130]	Proposed
Rotation (5°)	Baboon		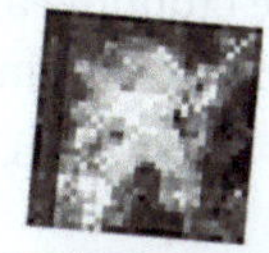	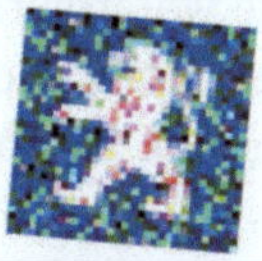
	Avion			
Rotation (30°)	Baboon			
	Avion			

Figure 6.15: The extracted watermarks via using different methods after rotation attack.

Attack	Images	Fan et al. [155]	Golea et al. [130]	Proposed
Cropping (25%)	Baboon	 0.75106	 0.73613	 0.83351
	Avion	 0.73106	 0.73613	 0.76354
Cropping (50%)	Baboon	 0.60674	 0.53112	 0.65661
	Avion	 0.50790	 0.53313	 0.57773

Figure 6.16: The extracted watermarks and NC values via using different methods after cropping attack.

does not exist the scrambling operation, the cropping position and size can directly affect the watermark in the cropping area. In Figure 6.16, the black area in the extracted watermark is the watermark information that is deleted as it is cropped in the area; therefore, the proposed watermarking method has higher anti-cropping robustness.

6.4.3 The Analysis of False-Positive Detection Problem

As mentioned earlier, the problem of false-positive detection arises in most SVD-based algorithms because only the singular values of the watermark are embedded into the host image, which will not be able to solve the copyright disputes. The solution to this problem is to embed all the information (not only the singular value) of the watermark and requires that the copyright owner must extract all the watermark information from the watermarked image (not only the singular value). In the method of this chapter, all the watermark information can be embedded into the host image and extracted from the host image, and hence, there is no false-positive detection problem.

6.5 Conclusion

This chapter presents a compensation optimization watermarking method based on SVD to embed color image watermark into color host image. Mainly perform SVD decomposition on 4×4 pixel block, and use the similarity between the entries in the second row of the first column and the one in the third row of the first column to embed and extract the watermark. Moreover, it can extract the embedded watermark from the watermarked image under variety of the attacks without original host image or the original watermark image. The experimental results show that the proposed watermarking method is both optimized in the invisibility and robustness of watermark.

7 The Color Image Watermarking Algorithm Based on Schur Decomposition

In this chapter, a blind dual-color image watermarking algorithm based on Schur decomposition is introduced. By analyzing the 4×4 unitary matrix U obtained by Schur decomposition, it is found that the entries between the second row the first column and the third row the first column have high similarity, which can embed watermark and extract watermark with blind manner. Experimental results show that the proposed algorithm has better robustness against most common attacks.

7.1 Introduction

The compensation optimization watermarking algorithm based on singular value decomposition (SVD) is proposed in Chapter 6, which is used to embed color image watermark into the color host image and has better invisibility. Although the algorithm can basically meet the requirements of the robustness of the watermark, its robustness is affected on a certain degree because the compensation operation reduces the differences of part of the compensated coefficient values, especially which cannot represent the original embedded relationship of the watermark information under attacks.

In recent years, in order to strengthen the digital copyright protection, researchers have put forward many watermarking algorithm based on SVD [141–147]. The successful application on SVD in digital watermarking suggests Schur decomposition also has the same function because the Schur decomposition is the main intermediate steps of SVD [153]. The computational complexity of Schur decomposition is $O(8N^3/3)$ and SVD is $O(11N^3)$. It is obvious that the number of computations required for Schur decomposition is less than one-third of SVD required. This relationship shows that the Schur decomposition will be more widely used in digital watermarking. Meanwhile, the Schur vector has good scaling invariance, which can improve the robustness of watermarking algorithm.

Therefore, on the basis of the introduced watermark embedding and extraction technology in Chapter 6, this chapter explores the use of Schur decomposition to embed color image watermark in color host image and blindly extract the watermark. Experimental results show that the proposed algorithm is invisible and very robust against the majority of common image processing attacks such as lossy compression, low-pass filtering, cropping, noise addition, blurring, rotation, scaling, sharpening, and so on. Comparison with the related SVD-based algorithm and spatial domain algorithm reveals that the proposed algorithm has better robustness under most attacks.

DOI 10.1515/9783110487732-007

7.2 The Schur Decomposition of Image Blocks

The matrix decomposition proposed by Schur in 1909 is a kind of typical unitary similarity transformation [160], which is an important tool in numerical linear algebra [153] and can be defined as follows:

If $A \in C^{N\times N}$, then there is a unitary matrix $U \in C^{N\times N}$, which makes

$$A = URU^T = U(D + N)U^T, \tag{7.1}$$

where D is a diagonal matrix containing the eigenvalues of matrix A, N is a strict upper triangular matrix, that is, $n_{ij} = 0, \forall i \geq j$, U^T indicates transposed matrix of U.

Schur decomposition has many features, here we only discuss two features in relation to their vector:

1. The invariance of the Schur vector subspace
   If $U = [u_1, u_2, \ldots, u_N]$ is the column block of the unitary matrix U, then u_i is called the Schur vector. To equal the column vector of both sides of $AU = UR$, we can see Schur vector meets the following relation equation:

$$Au_k = \lambda_k u_k + \sum_{i=1}^{k-1} N_{ik} u_i, \quad k = 1, 2, \ldots, n. \tag{7.2}$$

   It is concluded that the subspace $S_k = \text{span}\{u_1, u_2, \ldots, u_k\}$, $k = 1, 2, \ldots, n$, is invariant.
2. The scaling invariance of the Schur vector
   Multiply by the constant α on both sides of eq. (7.2), so

$$(\alpha A)u_k = (\alpha\lambda_k)u_k + \sum_{i=1}^{k-1} (\alpha N_{ik})u_i. \tag{7.3}$$

This shows that when the matrix A enlarges α times, Schur vector does not change, but its eigenvalue will be amplified α times. Using this feature, we can embed the watermark in Schur vector to resist scaling attacks.

Since the pixel values of the color image are between 0 and 255, the matrix entries from the color image should be nonnegative. In addition, the color image is divided into three layers of R, G, and B and each layer of image is gray-level image, without loss of generality, whose neighborhood pixel values are not obviously changed, and especially the size of pixel block in an image is smaller. Thus, there are two obvious features in the unitary matrix U of image block after Schur decomposition, that is, all the first column entries are of same signs and their values are very close. In this chapter, the 4×4 matrix A_1 is used to explain this feature:

$$A_1 = \begin{bmatrix} 185 & 186 & 187 & 188 \\ 184 & 185 & 186 & 187 \\ 184 & 184 & 185 & 186 \\ 184 & 185 & 186 & 186 \end{bmatrix}, \quad U_1 = \begin{bmatrix} 0.5027 & 0.0102 & 0.7729 & 0.3871 \\ 0.5000 & 0.7033 & -0.0850 & -0.4981 \\ 0.4980 & -0.7108 & -0.0680 & -0.4922 \\ 0.4993 & -0.0057 & -0.6252 & 0.5998 \end{bmatrix}. \quad (7.4)$$

In eq. (7.4), Schur decomposition of A_1 produces U_1 matrix, while the signs of the first column entries in U_1 matrix are the same. This is further verified by considering another sample matrix A_2 and the same case can be found in the U_2 matrix in eq. (7.5):

$$A_2 = \begin{bmatrix} 128 & 115 & 113 & 89 \\ 28 & 56 & 90 & 1 \\ 25 & 45 & 25 & 55 \\ 184 & 32 & 0 & 15 \end{bmatrix}, \quad U_2 = \begin{bmatrix} -0.7498 & 0.0888 & -0.5939 & -0.2778 \\ -0.2092 & 0.9071 & 0.3081 & 0.1960 \\ -0.2484 & -0.1166 & 0.6343 & -0.7227 \\ -0.5764 & -0.3945 & 0.3873 & 0.6017 \end{bmatrix}. \quad (7.5)$$

Moreover, this chapter divides some standard color images into 4 × 4 pixel blocks and gets the unitary matrices U after Schur decomposition. Suppose a matrix $U_{m,1}$ consists of mth row of the first column entry of each U matrix and another one $U_{n,1}$ consisting of nth row of the first column entry of each U matrix block are formed, respectively. NC between the two matrices $U_{m,1}$ and $U_{n,1}$ is computed by eq. (1.11) and listed in Table 7.1 with many standard test images. As can be seen from the table that the average value of NC $(U_{2,1}, U_{3,1})$ is 0.9672, which shows that the entries of the second row the first column and the third row the first column of the 4 × 4 matrix after Schur decomposition are very similar. Therefore, this stable similarity can be explored for embedding watermark and extracting watermark in the blind manner.

Table 7.1: The similarity of different elements in first column of U matrix after Schur decomposition (NC).

Image	NC($U_{1,1}$,$U_{2,1}$)	NC($U_{1,1}$,$U_{3,1}$)	NC($U_{1,1}$,$U_{4,1}$)	NC($U_{2,1}$,$U_{3,1}$)	NC($U_{2,1}$,$U_{4,1}$)	NC($U_{3,1}$,$U_{4,1}$)
Lena	0.9923	0.9875	0.9868	0.9969	0.9896	0.9933
Peppers	0.8868	0.8681	0.8656	0.9773	0.9475	0.9609
Avion	0.9248	0.9076	0.8989	0.9699	0.9461	0.9651
Baboon	0.9708	0.9593	0.9532	0.9793	0.9576	0.9714
Bear	0.8525	0.8300	0.8294	0.9528	0.9141	0.9348
TTU	0.9720	0.9554	0.9536	0.9908	0.9772	0.9858
Sailboat	0.8853	0.8718	0.8721	0.9851	0.9621	0.9736
Barbara	0.9440	0.9324	0.9290	0.9854	0.9740	0.9858
Couple	0.6090	0.5818	0.5900	0.8676	0.8516	0.8472
Average	0.8931	0.8771	0.8754	0.9672	0.9466	0.9575

7.3 The Color Image Watermarking Algorithm Based on Schur Decomposition

Without loss of generality, suppose the original host image H is 24-bit color image with size of $M \times M$ and the watermark image is 24-bit color image W with size of $N \times N$.

7.3.1 Watermark Embedding

The procedure of watermark embedding is shown in Figure 7.1. The detailed steps are introduced as follows:

1. Preprocess watermark image. The three-dimensional original watermark image W is firstly partitioned into three components R, G, and B by dimension-reduction treatment. And then, the component watermarks W_i ($i = 1, 2, 3$) are obtained, which presents the R, G, and B components, respectively. In order to enhance the security and robustness of the watermarking, each component watermark is permuted by Arnold transformation with the private key Ka and converted every pixel value into 8-bit binary sequence.
2. Block process the host image. The host image is divided into R, G, and B component images and each component image is partitioned into 4×4 nonoverlapping blocks.
3. Perform Schur decomposition on each block $H_{i,j}$ as eq. (7.6) to obtain the $U_{i,j}$ matrix of each block:

$$H_{i,j} = U_{i,j} S_{i,j} U_{i,j}^T. \tag{7.6}$$

4. Modify the entries of $u_{2,1}$ and $u_{3,1}$ in the $U_{i,j}$ matrix of each block to get the modified block $U'_{i,j}$ according to the watermark information $w_{i,j}$. According to the rules in eqs (7.7) and (7.8), the watermark $w_{i,j}$ is embedded by modifying the

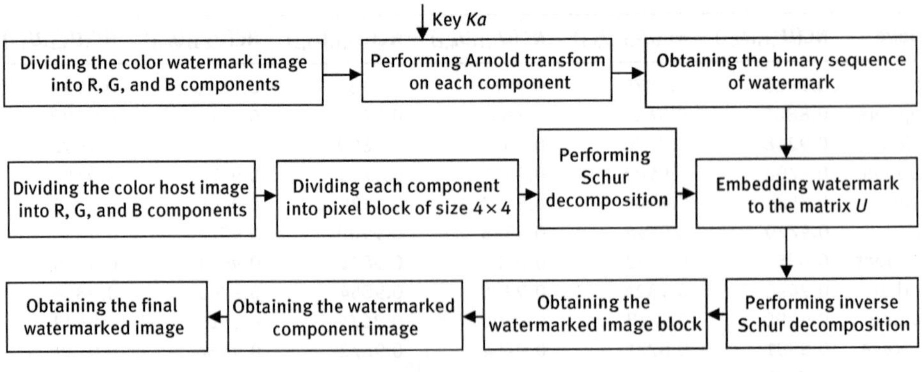

Figure 7.1: The diagram of watermark embedding.

relation between the second entry ($u_{2,1}$) and the third entry ($u_{3,1}$) of the first column:

$$\text{if } w_{i,j} = 1, \quad \begin{cases} u^*_{2,1} = \text{sign}(u_{2,1}) \times (U_{avg} + T/2) \\ u^*_{3,1} = \text{sign}(u_{3,1}) \times (U_{avg} - T/2) \end{cases}, \tag{7.7}$$

$$\text{if } w_{i,j} = 0, \quad \begin{cases} u^*_{2,1} = \text{sign}(u_{2,1}) \times (U_{avg} - T/2) \\ u^*_{3,1} = \text{sign}(u_{3,1}) \times (U_{avg} + T/2) \end{cases}, \tag{7.8}$$

where sign(x) presents the sign of x, $U_{avg} = (|u_{2,1}| + |u_{3,1}|)/2$, and $|x|$ denotes the absolute value of x

5. Obtain the watermarked image block by eq. (7.9).

$$H^*_{i,j} = U^*_{i,j} S_{i,j} U^{*T}_{i,j} \tag{7.9}$$

6. Repeat steps 3–5 until all watermark bits are embedded in the host image. Finally, recombine the watermarked R, G, and B components and obtain the watermarked image H^*

7.3.2 Watermark Extraction

The procedure of watermark extraction is shown in Figure 7.2. The detailed steps are presented as follows:

1. The watermarked image H^* is partitioned into R, G, and B component images, which are further partitioned into watermarked blocks with size of 4 × 4 pixels, respectively.
2. Perform Schur decomposition on the watermarked blocks $H'_{i,j}$ to get the $U'_{i,j}$ matrix.
3. The relation between the second ($u'_{2,1}$) and the third ($u'_{3,1}$) entries in the first column of the $U'_{i,j}$ matrix is used to extract the watermark information $w'_{i,j}$, as follows:

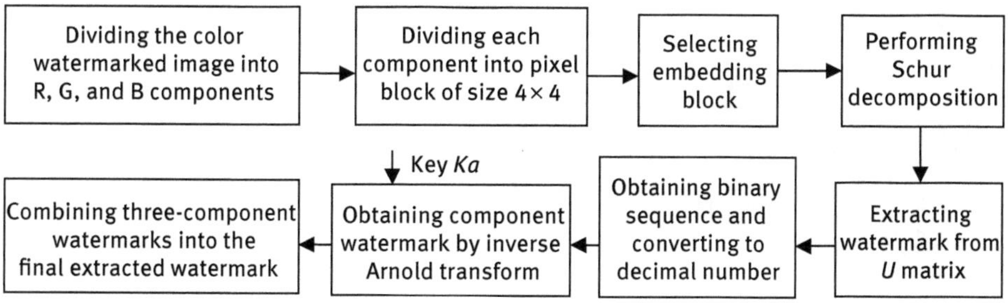

Figure 7.2: The diagram of watermark extraction.

$$w_{i,j}^{*} = \begin{cases} 0, & \text{if } |u_{2,1}^{*}| > |u_{3,1}^{*}| \\ 1, & \text{if } |u_{2,1}^{*}| \leq |u_{3,1}^{*}| \end{cases}. \tag{7.10}$$

4. Repeat steps 2 and 3 until all embedded image blocks are performed. These extracted bit values are partitioned into 8-bit groups and converted to decimal pixel values, then the inverse-Arnold transformation based on the private key *Ka* is executed and the extracted watermark of each component is reconstructed.
5. Reconstruct the final extracted watermark W^{*} from the extracted watermarks of the three components.

It is seen from this chapter that the proposed watermarking algorithm in the watermark extraction procedure can extract color image watermark from the watermarked image without the original watermark or the original host; thus, the watermarking algorithm can realize blind watermark extraction.

7.4 Algorithm Test and Result Analysis

In this experiment, four 24-bit color images with size of 512 × 512, as shown in Figure 7.3(a)–(d), that is, Lena, Avion, Peppers, and TTU, are selected as the original host images. Additionally, two 24-bit color images with size of 32 × 32, as shown in Figure 7.3(e) and 7.3(f), are used as original watermarks. At the same time, structural similarity index measurement (SSIM) is employed to evaluate the imperceptible capability of watermarking and normalized cross-correlation (NC) value is used to evaluate the watermark robustness.

Figure 7.3: Original host images: (a) Lena, (b) Avion, (c) peppers, and (d) TTU; original watermark images: (e) Peugeot logo and (f) 8-color image.

	Lena	Avion	Peppers	TTU
Watermarked image				
SSIM	0.94110	0.88758	0.96049	0.97669
Extracted watermark				
NC	1.00000	0.99860	1.00000	0.99799

Figure 7.4: The watermarked images (SSIM) and the extracted watermarks without any attacks (NC).

7.4.1 Test of the Watermark Invisibility

In order to evaluate the invisibility of watermarking, the watermark of Figure 7.3(e) is embedded in the host images of Figure 7.3(a) and 7.3(b), while the watermark of Figure 7.3(f) is also embedded in the host images of Figure 7.3(c) and 7.3(d), respectively. Figure 7.4 illustrates the watermarked color images and their SSIM values, which shows that the better visual quality of watermarked images can be obtained by the proposed scheme. In addition, the extracted watermarks can reflect the real appearance of the watermark and has high watermark visibility relative to the SVD algorithm in Chapter 6.

In Section 7.4.2, various attacks, such as Joint Photographic Expert Group (JPEG) compression, JPEG 2000 compression, cropping, adding noise, scaling, low-pass filtering, median filtering, rotation, blurring, and so on, are performed on the watermarked images Lean and Avion to test the robustness of the proposed method and compared the proposed method with the related work.

7.4.2 Test of the Watermark Robustness

JPEG compression attack is one of the common attacks that should be verified in watermarking algorithm. Figure 7.5 shows the experimental results when the compression factors are 30 and 90, respectively. Compared with the method in Ref. [130], the proposed scheme achieves a better robustness against the JPEG compression.

JPEG 2000 was developed by the JPEG with the aim of improving the properties of the JPEG standard. The watermarked images are also performed by JPEG 2000

Attack	Image	Golea et al. [130]		Proposed	
		NC	Extracted watermark	Extracted watermark	NC
JPEG 30	Lena	0.65311			0.82130
	Avion	0.82043			0.84098
JPEG 90	Lena	0.97719			0.99313
	Avion	0.97595			0.99416

Figure 7.5: The extracted watermarks and NC values via using different methods after JPEG compression attack.

compression with the compression ratio ranging from 1 to 10 increasing in steps of 1. Figure 7.6 presents the NC values and the visual perception results with compression ratios of 5 and 10, respectively.

Figure 7.7 shows the result of extracted watermark images after salt and peppers noise attacks whose noise intensities are 2 % and 10 %, respectively. It is evident from these results that the proposed method has better robustness against salt and peppers noise attacks than the method in Ref. [130].

In addition, the watermarked images will be attacked by Gaussian noise whose mean values are 0.1 and 0.3, respectively. The extracted watermarks and the NC values are shown in Figure 7.8.

Figure 7.9 shows the results for median filtering attack. As shown in Figure 7.9, although the visual quality of the two methods is not good, the proposed method is relatively better than the method in Ref. [130]. The main reason of poor result is that the watermark embedding and extraction of this chapter is on the basis of pixel division of 4×4 blocks, while median filter sizes were set to 2×2 and 3×3, respectively.

In this experiment, the Butterworth low-pass filters are used to attack the watermarked images with cutoff frequency 100, and order 1, 3, respectively. Figure 7.10 shows the NC values and the visual perception results, respectively.

Attack	Image	Golea et al. [130]		Proposed	
		NC	Extracted watermark	Extracted watermark	NC
JPEG2000 (5:1)	Lena	0.93898			0.99096
	Avion	0.93748			0.99526
JPEG2000 (10:1)	Lena	0.80712			0.81363
	Avion	0.79507		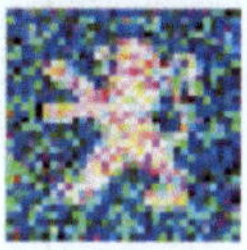	0.90744

Figure 7.6: The extracted watermarks and NC values via using different methods after JPEG 2000 compression attack.

Attack	Image	Golea et al. [130]		Proposed	
		NC	Extracted watermark	Extracted watermark	NC
Salt & Peppers noise (0.02)	Lena	0.56979			0.95413
	Avion	0.52759			0.95382
Salt & Peppers noise (0.10)	Lena	0.21502		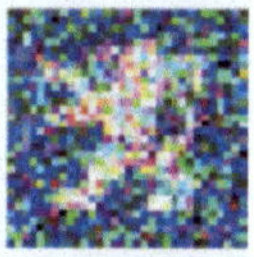	0.80093
	Avion	0.21075			0.80176

Figure 7.7: The extracted watermarks and NC values via using different methods after salt and peppers noising attack.

Attack	Image	Goleaet al. [130]		Proposed	
		NC	Extracted watermark	Extracted watermark	NC
Gaussian noise (0.1)	Lena	0.86001			0.96598
	Avion	0.81877			0.97281
Gaussian noise (0.3)	Lena	0.81641			0.91618
	Avion	0.65114			0.50636

Figure 7.8: The extracted watermarks and NC values via using different methods after Gaussian noising attack.

Attack	Image	Golea et al. [130]		Proposed	
		NC	Extracted watermark	Extracted watermark	NC
Median Filter (2×2)	Lena	0.65192			0.87088
	Avion	0.67150			0.91028
Median Filter (3×3)	Lena	0.50744			0.54118
	Avion	0.53405			0.50191

Figure 7.9: The extracted watermarks and NC values via using different methods after median filtering attack.

Attack	Image	Golea et al. [130]		Proposed	
		NC	Extracted watermark	Extracted watermark	NC
Low-pass Filter (100,1)	Lena	0.54766			0.97150
	Avion	0.58553			0.92275
Low-pass Filter (100,3)	Lena	0.40915			0.88517
	Avion	0.46500			0.83498

Figure 7.10: The extracted watermarks and NC values via using different methods after low-pass filtering attack.

Attack	Image	Golea et al. [130]		Proposed	
		NC	Extracted watermark	Extracted watermark	NC
Sharpening (0.2)	Lena	0.84806			0.99907
	Avion	0.85027			0.98893
Sharpening (1.0)	Lena	0.78075			0.80712
	Avion	0.62558			0.98218

Figure 7.11: The extracted watermarks and NC values via using different methods after sharpening attack.

Attack	Image	Golea et al. [130]		Proposed	
		NC	Extracted watermark	Extracted watermark	NC
Scaling (4)	Lena	0.83852			0.99167
	Avion	0.86885			0.96679
Scaling (1/4)	Lena	0.56979			0.90411
	Avion	0.6146			0.84933

Figure 7.12: The extracted watermarks and NC values via using different methods after scaling attack.

Attack	Image	Golea et al. [130]		Proposed	
		NC	Extracted watermark	Extracted watermark	NC
Blurring (0.2)	Lena	1.00000			1.00000
	Avion	0.57185			0.98893
Blurring (1.0)	Lena	0.27017			0.88549
	Avion	0.17850	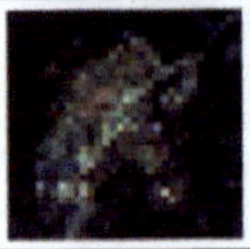		0.80100

Figure 7.13: The extracted watermarks and NC values via using different methods after blurring attack.

Figure 7.11 shows the results of sharpening attack. During the process of sharpening, the radii are 0.2 and 1.0, respectively. It is proved that the proposed method outperforms the method in Ref. [130].

Two scaling operations of 400 % and 25 % are utilized to deteriorate the watermarked image in this chapter. Figure 7.12 shows the quantitative results and the visual perception results with the case of scaling. The proposed algorithm of this chapter has better capability to resist scaling attack because Schur vector has better scale invariance.

In addition, two cases are simulated to degrade the two watermarked images. The first case with the radius is 0.2, while the second case with the radius is 1.0. Figure 7.13 lists the results of visual comparison and the NC values.

Figure 7.14 shows the results of two kinds of rotation attack. One involves rotating the watermarked image to the right by 5°. The other experiment involves rotating the watermarked image to the right by 30°. In each attack, the images are first rotated with a certain number of degrees clockwise and rotated with the same number of degrees counterclockwise, then cropped and scaled to gain the 512 × 512 images that are used to extract the watermark.

Figure 7.15 shows the results of two kinds of cropping attack. The first case is cropped by 25 %, while the second one is cropped by 50 %. Because the watermark

Attack	Image	Golea et al. [130]	Proposed
		Extracted watermark	Extracted watermark
Rotation 5°	Lena	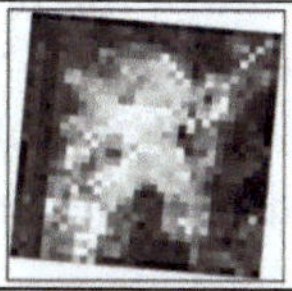	
	Avion		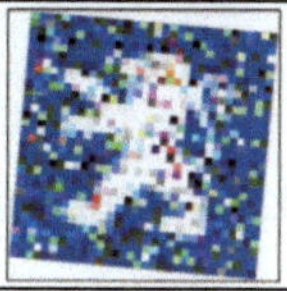
Rotation 30°	Lena	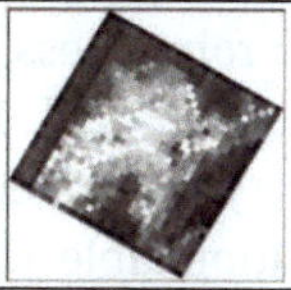	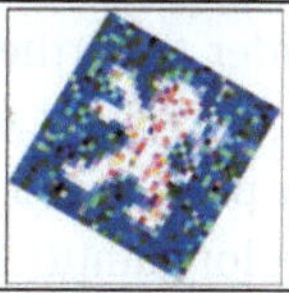
	Avion	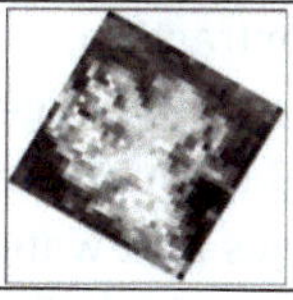	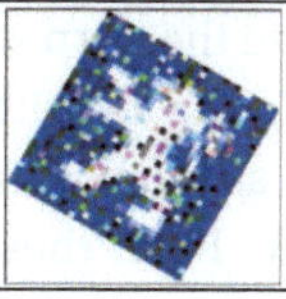

Figure 7.14: The extracted watermarks via using different methods after rotation attack.

Attack	Image	Golea et al. [130]		Proposed	
		NC	Extracted watermark	Extracted watermark	NC
Cropping (25%)	Lena	0.73804			0.89679
	Avion	0.73613			0.78327
Cropping (50%)	Lena	0.53112			0.64674
	Avion	0.53313			0.56319

Figure 7.15: The extracted watermarks and NC values via using different methods after cropping attack.

used in Ref. [130] is not permuted, the position and size of cropping can fully affect the watermark in the cropped region. There is a black region in the extracted watermark, which means the watermark information in this region is fully deleted by cropping attack. Hence, the quality of the proposed method outperforms the method in Ref. [130].

7.4.3 Comparison with the Spatial Domain Algorithm

In order to further prove the robustness, the proposed methods are also compared with the spatial domain algorithm [83]. Here, the color quantization index of each color pixel is modified to carry the watermark in the embedding process; however, the color gamut and quantization table are required in the extraction, which does not realize the purpose of blind extraction.

In the algorithm [83], two color images, as shown in Figure 7.3(c) and 7.3(d), were used as the host image, and 8-color image of Figure 7.3(f) was taken as the watermark image. Here, these host images and watermark image are also used to carry out the experiment under same attack styles in Ref. [83]. These results, as shown in Figure 7.16, reveal that the proposed algorithm has better robustness. This is because the changed

Host image	Attack method	Chou et al. [83]		Proposed method	
		NC	Extracted watermark	NC	Extracted watermark
Peppers	Low-pass filtering	0.539		0.94085	
	Crop50%	0.553		0.92000	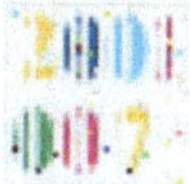
	Scaling1/4	0.536		0.89943	
	Scaling4	0.851		0.97513	
	Rotation 30	–		–	
	JPEG 12:1	0.439		0.97886	
	JPEG 27.5:1	0.343	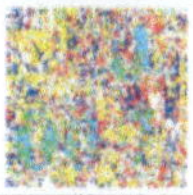	0.93404	
TTU	Low-pass filtering	0.423		0.89168	
	Gaussian noise 4	0.982		0.93892	
	Gaussian noises 25	0.360		0.75349	
	Median Filter (3×3)	0.170		0.34617	

Figure 7.16: The extracted watermarks via using different methods after different attacks.

color values for various attacks have directly affected the mapping relation between the original color value and the color table, which results in the degraded quality of extracted watermark in the algorithm [83].

7.5 Conclusion

In this paper, we have proposed an algorithm based on Schur decomposition for embedding the color image watermark into the color host image, which may serve as a measurable way to protect the copyright of color image. It successfully employs the strong relation between the second row of the first column entry and the third row of the first column entry of 4×4 U matrix in the Schur decomposition to embed and extract watermark. The embedded watermark can be extracted from the different attacked images without resorting to the original host image or original watermark. Experimental results have shown that this algorithm not only guarantees the invisibility of watermarking but also has strong robustness during the operation of common image processing.

8 The Color Image Watermarking Algorithm Based on QR Decomposition

This chapter proposes an efficient color image blind watermarking algorithm based on QR decomposition. First, the color host image is divided into 4×4 nonoverlapping pixel blocks. Then, each selected pixel block is decomposed by QR decomposition and the entry in the first row of the fourth column of the matrix R is quantified for embedding the watermark information. In the extraction procedure, the watermark can be extracted from the watermarked image without the original host image or the original watermark image. Experimental results show that the scheme not only meets the basic requirements of watermark properties, but also has very high execution efficiency for hardware converting and practicability of the algorithms.

8.1 Introduction

Some former chapters have researched the embedding and extraction problems of color image digital watermark, respectively, from the aspects of watermark capacity, invisibility, and robustness, which is, respectively, suitable for different occasions in different requests. In order to facilitate practical hardware, we need to design a fast and effective watermarking algorithm.

In Chapter 7, we analyzed that the time complexity of singular value decomposition (SVD) or Schur decomposition are both $O(N^3)$, and found that the time complexity of QR decomposition is $O(N^2)$, by further study, whose feature has accelerated the application of QR decomposition in the digital watermarking [161–163]. In the past few years, digital watermarking algorithm based on QR decomposition appeared. Yasha et al. [164] proposed to embed a watermark bit in all entries of the first row of R matrix after each 8×8 block was decomposed by QR decomposition, in which the watermark was 88×88 binary image. By modifying the entry of the matrix Q after QR decomposition, method in Ref. [165] embedded a 32×32 binary image into 512×512 host image. The common features of these two kinds of watermarking technology all use binary image as watermark image. The watermark information of the color image is 24 times more than that of the binary image with the same size when the color image is used as watermark; thus, the methods theoretically described in Refs [164, 165] cannot meet the requirements of the color image as watermark.

According to the above discussion, this chapter put forward a kind of efficient double-color image watermarking algorithm based on QR decomposition. By theoretical analysis and experimental analysis, it is found that the first row of the fourth column entry of the R matrix can be quantified for embedding watermark. On the basis of the analysis of many experimental data, the quantization step is selected for keeping the trade-off between the robustness and imperceptibility of the embedded watermark. In addition, the proposed watermarking algorithm can achieve

DOI 10.1515/9783110487732-008

the purpose of blind extraction. The simulation data shows that not only the algorithm satisfies the watermark invisibility and the strong robustness, but also the performance efficiency of the algorithm has been significantly increased.

8.2 The QR Decomposition of Image Block

Suppose A is a nonsingular matrix with size of $N \times N$, its QR decomposition can be represented as

$$[Q, R] = qr(A), \tag{8.1}$$

where Q is an $N \times N$ matrix with standard orthogonal vector, whose column vectors are obtained from columns of A by Gram–Schmidt orthogonalization process; R is an $N \times N$ upper triangular matrix, and set $A = [a_1, a_2, \ldots, a_n]$, $Q = [q_1, q_2, \ldots, q_n]$, where a_i and q_i, respectively, are column vectors, $i = 1, 2, \ldots, n$, so the R matrix can be obtained by the following equation:

$$R = \begin{bmatrix} \langle a_1, q_1 \rangle & \langle a_2, q_1 \rangle & \cdots & \langle a_n, q_1 \rangle \\ 0 & \langle a_2, q_2 \rangle & \cdots & \langle a_n, q_2 \rangle \\ \vdots & \vdots & \ddots & \vdots \\ 0 & 0 & 0 & \langle a_n, q_n \rangle \end{bmatrix}, \tag{8.2}$$

where $\langle a_i, q_i \rangle$ is the inner product of vectors a_i and q_i.

One of the important properties of R matrix after QR decomposition is that when the columns of A have correlation with each other, the absolute values of the entries in the first row of R matrix are probably greater than those in other rows [163].

The condition and probability that satisfy the property will be analyzed as follows.

8.2.1 Condition Analysis

Without loss of generality, one 2×2 matrix $A = \begin{bmatrix} a & c \\ b & d \end{bmatrix}$ is used to verify the feature mentioned above. Meanwhile, according to the feature of image pixel, the value of each entry in matrix A is in the range of 0–255. According to the definition of QR decomposition, the QR decomposed results of matrix A can be obtained by

$$Q = \frac{1}{\sqrt{a^2 + b^2}} \begin{bmatrix} a & -b \\ b & a \end{bmatrix}, \tag{8.3}$$

$$R = \begin{bmatrix} r_{11} & r_{12} \\ r_{21} & r_{22} \end{bmatrix} = \frac{1}{\sqrt{a^2 + b^2}} \begin{bmatrix} a^2 + b^2 & ac + bd \\ 0 & ad - bc \end{bmatrix}. \tag{8.4}$$

1. The analysis of the relationship between r_{11} and r_{21}
   Since matrix A is nonsingular matrix, that is, $|A| \neq 0$, then

$$ad \neq bc. \tag{8.5}$$

   It is known from eq. (8.5) that pixel $a \neq 0$ and $b \neq 0$ simultaneously, and at least one value is between 1 and 255; hence

$$a^2 + b^2 \geq 1. \tag{8.6}$$

   From eqs (8.4) and (8.6), we can find that

$$r_{11} > r_{21}. \tag{8.7}$$

2. The analysis of the relationship between r_{12} and r_{22}
   It is divided into three conditions to be discussed:
   (1) $a = b \neq 0$
   It is known from eq. (8.4) that

$$|r_{12}| = \frac{ac + bd}{\sqrt{a^2 + b^2}} = \frac{c + d}{\sqrt{2}}, |r_{22}| = \frac{|ad - bc|}{\sqrt{a^2 + b^2}} = \frac{|d - c|}{\sqrt{2}}. \tag{8.8}$$

   The probability of '$a = b \neq 0$ and $c = 0$' is

$$P(a = b \neq 0, c = 0) = \frac{1}{255} \times \frac{1}{255} \times \frac{1}{256} = 6.007 \times 10^{-8}, \tag{8.9}$$

   which shows that this case has lower probability.
   When $a = b \neq 0$ and $c \neq 0$, then

$$c + d > |d - c|. \tag{8.10}$$

   As can be seen from eqs (8.8), (8.9), and (8.10), $|r_{12}|$ is almost greater than $|r_{22}|$ when $a = b \neq 0$.
   (2) $a > b \geq 0$
   If $|r_{12}| > |r_{22}|$, then the following equation is obtained:

$$\frac{ac + bd}{\sqrt{a^2 + b^2}} > \frac{|ad - bc|}{\sqrt{a^2 + b^2}} \Rightarrow ac + bd > |ad - bc|. \tag{8.11}$$

   If $ad - bc > 0$, it follows from eq. (8.11) that

$$ac+bd > ad-bc \Rightarrow ac+bc > ad-bd \Rightarrow (a+b)c > (a-b)d \Rightarrow d < \frac{a + b}{a - b}c. \tag{8.12}$$

If $ad - bc < 0$, eq. (8.11) can be further deducted as

$$ac + bd > bc - ad \Rightarrow ac - bc > -ad - bd \Rightarrow (a - b)c > -(a + b)d$$
$$\Rightarrow d > \frac{b - a}{a + b}c. \quad (8.13)$$

It is known from eqs (8.12) and (8.13) that if $a > b \geq 0$, then $|r_{12}| > |r_{22}|$ will be established when $\frac{b-a}{a+b}c < d < \frac{a+b}{a-b}c$.

(3) $0 \leq a < b$

If $|r_{12}| > |r_{22}|$, we can obtain the following equation:

$$\frac{ac + bd}{\sqrt{a^2 + b^2}} > \frac{|ad - bc|}{\sqrt{a^2 + b^2}} \Rightarrow ac + bd > |ad - bc|\,. \quad (8.14)$$

If $ad - bc > 0$, eq. (8.14) can be further deduced as

$$ac + bd > ad - bc \Rightarrow ac + bc > ad - bd \Rightarrow (a + b)c > (a - b)d \Rightarrow d > \frac{a + b}{a - b}c. \quad (8.15)$$

If $ad - bc < 0$, eq. (8.14) can be further deduced as

$$ac + bd > bc - ad \Rightarrow ad + bd > bc - ac \Rightarrow (a + b)d > (b - a)c \Rightarrow d > \frac{b - a}{a + b}c. \quad (8.16)$$

It is known from eqs (8.15) and (8.16) that if $0 \leq a < b$, then $|r_{12}| > |r_{22}|$will be established when $d > \frac{a+b}{a-b}c$ and $d > \frac{b-a}{a+b}c$.

8.2.2 The Analysis of the Probability

According to the above condition analysis, Figure 8.1(a) and 8.1(b), respectively, shows the condition area in the Cartesian coordinate system under $a > b \geq 0$ and $0 \leq a < b$, respectively.

As shown in Figure 8.1(a), when $a > b \geq 0$ and the values of (c, d) are located in the shadow range, the probability of $|r_{12}| > |r_{22}|$ is given by

$$P((|r_{12}| > |r_{22}|)|(a > b \geq 0)) = \frac{S_1}{S_{all}} = \frac{\left(\left(1 - \frac{a-b}{a+b}\right) l + l\right) \times l}{2l^2} = 1 - \frac{a - b}{2(a + b)}, \quad (8.17)$$

where $P((|r_{12}| > |r_{22}|)|(a > b \geq 0))$ represents the probability of $|r_{12}| > |r_{22}|$ in the condition $a > b \geq 0$, S_1 represents the area of the shadow range, and S_{all} is the total area of the permitted range of (c, d).

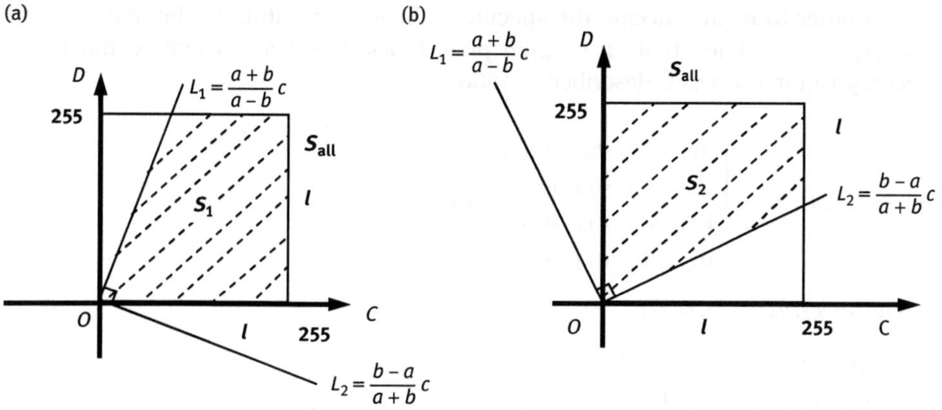

Figure 8.1: The illustration of condition area: (a) $a > b \geq 0$ and (b) $0 \leq a < b$.

As shown in Figure 8.1(b), when $0 \leq a < b$ and the values of (c, d) are located in the shadow range, the probability of $|r_{12}| > |r_{22}|$ is given by

$$P((|r_{12}| > |r_{22}|)|(0 \leq a < b)) = \frac{S_2}{S_{all}} = \frac{\left(\left(1 - \frac{b-a}{a+b}\right) l + l\right) \times l}{2l^2} = 1 - \frac{b-a}{2(a+b)}, \tag{8.18}$$

where $P((|r_{12}| > |r_{22}|)|(0 \leq a < b))$ represents the probability of $|r_{12}| > |r_{22}|$ in the condition $0 \leq a < b$, S_2 denotes the area of shadow range, and S_{all} is the total area of permitted range of (c, d).

In summary, when $a \neq b$, the probability of $|r_{12}| > |r_{22}|$ is obtained by

$$\begin{aligned} P(|r_{12}| > |r_{22}|) &= P((|r_{12}| > |r_{22}|)|(a > b \geq 0)) \times P(a > b \geq 0) \\ &\quad + P((|r_{12}| > |r_{22}|)|(0 \leq a < b)) \times P(0 \leq a < b)) \\ &= \frac{1}{2} \times (P((|r_{12}| > |r_{22}|)|(a > b \geq 0)) + P((|r_{12}| > |r_{22}|)|(0 \leq a < b))) \\ &= 1 - \frac{|b-a|}{2(a+b)}. \end{aligned} \tag{8.19}$$

As given in eq. (8.19) that the closer the values of a and b are, the bigger the value of $P(|r_{12}| > |r_{22}|)$ is, and the greater the value of $|r_{12}| > |r_{22}|$ is. Due to the correlation of the image pixel value, the difference between the neighborhood pixels is not obvious, so the probability value is generally large.

It is noted by the above discussion that the absolute values of the first row entries in R matrix obtained by QR decomposition are likely to be greater than those of the corresponding entries in other rows. Because the matrix entry with larger value allows a greater modified range, it is suitable for watermarking embedded, then which entry in 4×4 R matrix is more suitable for embedding watermark?

In order to further decide the specific embedding position in the first row of R matrix, it is assumed that an original pixel block A is a 4×4 matrix and its QR decomposition process is described as follows:

$$A = [a_1, a_2, a_3, a_4] = \begin{bmatrix} a_{1,1} & a_{1,2} & a_{1,3} & a_{1,4} \\ a_{2,1} & a_{2,2} & a_{2,3} & a_{2,4} \\ a_{3,1} & a_{3,2} & a_{3,3} & a_{3,4} \\ a_{4,1} & a_{4,2} & a_{4,3} & a_{4,4} \end{bmatrix} = QR$$

$$= [q_1, q_2, q_3, q_4]\,[r_1, r_2, r_3, r_4] \tag{8.20}$$

$$= \begin{bmatrix} q_{1,1} & q_{1,2} & q_{1,3} & q_{1,4} \\ q_{2,1} & q_{2,2} & q_{2,3} & q_{2,4} \\ q_{3,1} & q_{3,2} & q_{3,3} & q_{3,4} \\ q_{4,1} & q_{4,2} & q_{4,3} & q_{4,4} \end{bmatrix} \begin{bmatrix} r_{1,1} & r_{1,2} & r_{1,3} & r_{1,4} \\ 0 & r_{2,2} & r_{2,3} & r_{2,4} \\ 0 & 0 & r_{3,3} & r_{3,4} \\ 0 & 0 & 0 & r_{4,4} \end{bmatrix}$$

$$= \begin{bmatrix} q_{1,1}r_{1,1} & q_{1,1}r_{1,2} + q_{1,2}r_{2,2} & q_{1,1}r_{1,3} + q_{1,2}r_{2,3} + q_{1,3}r_{3,3} & q_{1,1}r_{1,4} + q_{1,2}r_{2,4} + q_{1,3}r_{3,4} + q_{1,4}r_{4,4} \\ q_{2,1}r_{1,1} & q_{2,1}r_{1,2} + q_{2,2}r_{2,2} & q_{2,1}r_{1,3} + q_{2,2}r_{2,3} + q_{2,3}r_{3,3} & q_{2,1}r_{1,4} + q_{2,2}r_{2,4} + q_{2,3}r_{3,4} + q_{2,4}r_{4,4} \\ q_{3,1}r_{1,1} & q_{3,1}r_{1,2} + q_{3,2}r_{2,2} & q_{3,1}r_{1,3} + q_{3,2}r_{2,3} + q_{3,3}r_{3,3} & q_{3,1}r_{1,4} + q_{3,2}r_{2,4} + q_{3,3}r_{3,4} + q_{3,4}r_{4,4} \\ q_{4,1}r_{1,1} & q_{4,1}r_{1,2} + q_{4,2}r_{2,2} & q_{4,1}r_{1,3} + q_{4,2}r_{2,3} + q_{4,3}r_{3,3} & q_{4,1}r_{1,4} + q_{4,2}r_{2,4} + q_{4,3}r_{3,4} + q_{4,4}r_{4,4} \end{bmatrix}.$$

It can be seen from eq. (8.20) that the entry $a_{1,1}$ is equal to $q_{1,1}r_{1,1}$, which means that the change of $r_{1,1}$ will directly affect the $a_{1,1}$ and will change the pixel value to affect the invisibility of the watermark, but the change of $r_{1,4}$ will have indirect effect on $a_{1,4}$. Thus, $r_{1,4}$ is the best entry for embedding watermark, which is further proved in the following experiment.

In the experiment, the different entries in the first row of R are modified for embedding watermark, respectively, and then the watermark is extracted from the attacked images that are attacked by 10 different types. The bigger the normalized correlation (NC) value is, the better the extracted watermark is, and the more suitable for embedding watermark the position is. As can be readily seen from Figure 8.2, it has better performance in embedding watermark into the entry $r_{1,4}$.

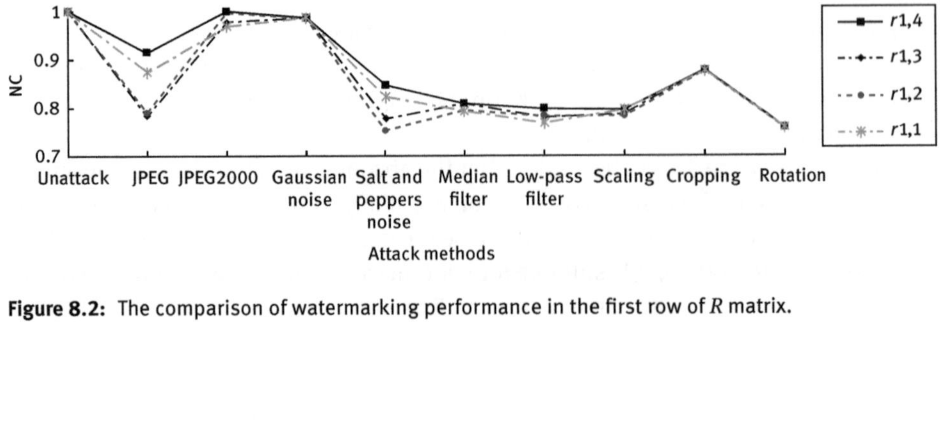

Figure 8.2: The comparison of watermarking performance in the first row of R matrix.

8.3 Color Image Watermark Algorithm Based on QR Decomposition

In this section, it will propose a kind of color watermark technology based on QR decomposition, and its aim is to embed a 24-bit color image W of size $N \times N$ into a 24-bit color original host image H of size $M \times M$ and achieve the aim of blind extraction. The detailed specific watermark embedding algorithm is described as follows:

8.3.1 Watermark Embedding

The process of watermark embedding is shown in Figure 8.3, and the specific steps are described as follows:

8.3.1.1 Preprocessing on Color Image Watermark

First, three-dimensional original color watermark image W is divided into three two-dimensional components R, G, and B by dimension-reduction treatment. In order to improve the security and robustness of the watermarking, each component watermark is permuted by Arnold random numbers with private key $KA_i(i = 1, 2, 3)$ and each pixel value is converted to 8-bit binary sequence. Finally, combining all 8-bit binary sequence to form the binary component watermark W_i $(i = 1, 2, 3)$ respectively represents R, G, and B color.

8.3.1.2 Block Processing of the Host Image

The original host image is divided into R, G, and B three-component images H_i $(i = 1, 2, 3)$, and each component image is divided into 4×4 nonoverlapping blocks.

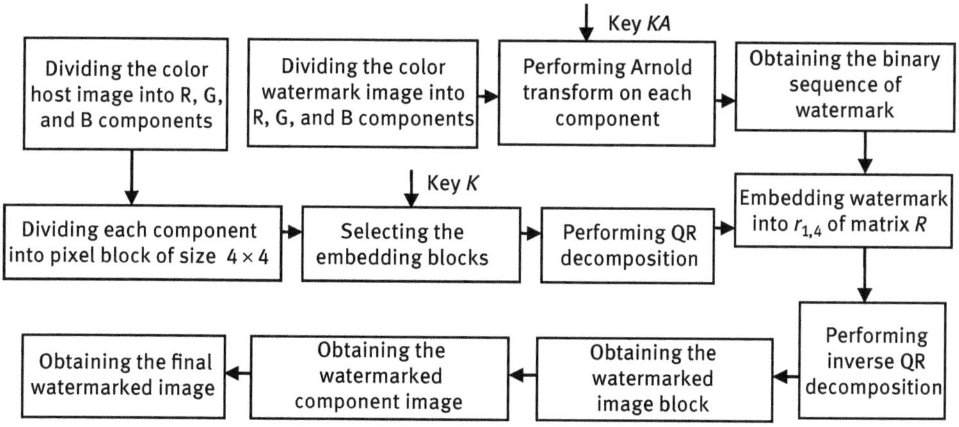

Figure 8.3: The diagram of watermark embedding.

8.3.1.3 Selecting the Embedding Blocks

In order to ensure the security of the watermark, the MD5-Hash pseudorandom replacement algorithm based on private key K_i ($i = 1, 2, 3$) is used to select the noncollision embedding pixel block in image H_i ($i = 1, 2, 3$).

8.3.1.4 QR Decomposition

Each selected 4×4 block is decomposed by QR decomposition according to eq. (8.1).

8.3.1.5 Embedding Watermark

The process of modifying $r_{1,4}$ to embed watermark information w is given as follows:

1. Decide the modifying magnitudes T_1 and T_2 according to the watermark information w:

$$\text{if } w = 1, \quad \begin{cases} T_1 = 0.5\Delta \\ T_2 = -1.5\Delta \end{cases}, \tag{8.21}$$

$$\text{if } w = 0, \quad \begin{cases} T_1 = -0.5\Delta \\ T_2 = 1.5\Delta \end{cases}. \tag{8.22}$$

2. The possible quantization results C_1 and C_2 can be computed by the modifying magnitudes T_1 and T_2:

$$C_1 = 2k\Delta + T_1, \tag{8.23}$$

$$C_2 = 2k\Delta + T_2, \tag{8.24}$$

   where $k = \text{floor}(\text{ceil}(r_{1,4}/\Delta)/2)$, floor($x$) is the biggest integer that is not more than x, and ceil(x) is the smallest integer that is not less than x.

3. Calculate the value $r'_{1,4}$ of the embedding watermark by the following condition:

$$r^*_{1,4} = \begin{cases} C_2, & \text{if} \quad \text{abs}(r_{1,4} - C_2) < \text{abs}(r_{1,4} - C_1) \\ C_1, & \text{otherwise} \end{cases}, \tag{8.25}$$

   where abs(x) denotes the absolute value of x.

8.3.1.6 Inverse QR Decomposition

Replace $r_{1,4}$ with $r^*_{1,4}$, and then apply inverse QR decomposition to obtain the watermarked image block as follows:

$$A^* = Q \times R^*. \tag{8.26}$$

8.3.1.7 Looping

Repeat steps 8.3.1.4–8.3.1.6 until all the watermark information is embedded in the host image. Finally, the watermarked R, G, and B components are reconstructed to obtain the watermarked image H^*.

8.3.2 Watermark Extraction

In this watermark extraction algorithm, the original host image or watermark image is not needed in the procedure. The process is shown in Figure 8.4, and the detailed steps of the watermark extraction procedure are presented as follows.

8.3.2.1 Preprocessing on the Watermarked Image

The watermarked image H^* is divided into R, G, and B three-component images, which are further divided into watermarked blocks with size of 4×4 pixels, respectively.

8.3.2.2 Selecting the Watermarked Block

The MD5-Hash pseudorandom replacement algorithm based on private key K_i $(i = 1, 2, 3)$ is used to select the watermarked pixel blocks in component image.

8.3.2.3 QR Decomposition

According to eq. (8.1), each watermarked block is decomposed by QR decomposition and the matrix R^* is obtained.

8.3.2.4 Watermark Extraction

The first row fourth column entry $r^*_{1,4}$ of matrix R^* is used to extract the watermark information w^*, as follows:

$$w^* = \mathrm{mod}(\mathrm{ceil}(r^*_{1,4}/\Delta), 2), \tag{8.27}$$

where mod(.) is the modulo operation.

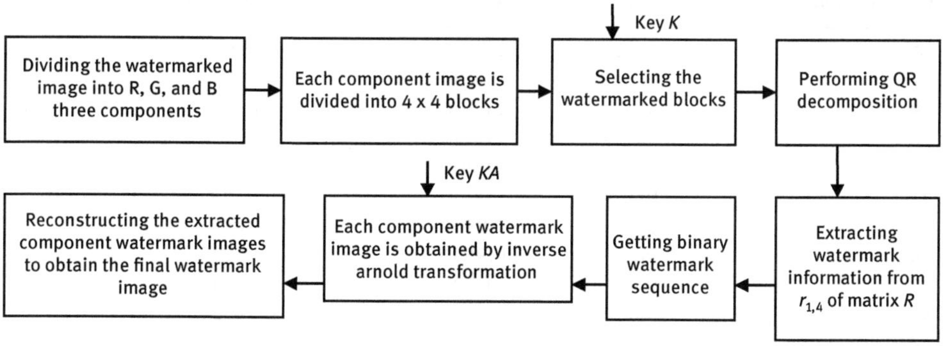

Figure 8.4: The diagram of watermark extraction.

8.3.2.5 Looping

The above steps 8.3.2.2–8.3.2.4 are repeated until all embedded image blocks are performed. These extracted watermark bit values are partitioned into 8-bit groups and converted to decimal pixel values.

8.3.2.6 Reconstruction

Each component watermark is transformed by the inverse Arnold transformation based on the private key KA_i $(i = 1, 2, 3)$. Then the final extracted watermark W^* is reconstructed from the extracted watermarks of the three components.

8.4 Experimental Results and Discussion

In this experiment, four 24-bit 512 × 512-sized color images (Lena, Avion, Peppers, and TTU) in Figure 7.3(a)–7.3(d) are used as the original host images, and two 24-bit color images with size of 32 × 32 in Figure 7.3(e) and 7.3(f) are used as the original host watermarks. Meanwhile, the structural similarity index measurement (SSIM) is used to evaluate the invisibility of the watermark and the normalized cross-correlation (NC) is used to evaluate the robustness of the watermark.

8.4.1 The Selection of Quantization Step

In order to ensure the invisibility and robustness of the watermark, this chapter selects the suitable quantization step Δ by using plenty of experiments. From Table 8.1 it can be seen that the bigger the quantifying step Δ is, the worse the invisibility of the watermark is, but the stronger the robustness of the watermark is. Considering the balance of both, quantization step Δ is set to 38.

Table 8.1: The comparison of watermarking performance with different steps.

Quantization step	SSIM	NC (nonattack)	Average NC (different attacks)
10	0.99673	1.0000	0.7589
14	0.99369	1.0000	0.8206
18	0.98932	1.0000	0.8525
20	0.98727	1.0000	0.8675
24	0.98136	1.0000	0.8899
28	0.97557	1.0000	0.9084
30	0.97269	0.9998	0.9158
34	0.96475	0.9998	0.9259
38	0.95686	1.0000	0.9433
40	0.95277	0.9999	0.9463
44	0.94422	0.9999	0.9567
48	0.93651	0.9998	0.9626

8.4.2 Test of the Watermark Invisibility

In order to verify the watermark invisibility of the algorithm in this chapter, many different host images and watermark images are used to compare with different algorithms. Besides the quantitative results in terms of SSIM and NC, the experiment also provides the visual comparison results. During the experiment, the watermark shown in Figure 7.3(e) is embedded, respectively, into the host images shown in Figure 7.3(a) and 7.3(b); meanwhile, the watermark shown in Figure 7.3(f) is embedded into the host images shown in Figure 7.3(c) and 7.3(d).

Figure 8.5 not only gives the watermarked color images and their SSIM values, but also shows the extracted watermark under the no-attacked condition. By comparison, it can be seen that the method that proposed by Song et al. [165] based on QR decomposition cannot better extract the watermark and the watermarked host image arises obvious changes, which cannot meet the requirement of watermark invisibility, so the algorithm does not suit for embedding color image watermark into color host image. Comparatively, the algorithm based on SVD in Ref. [130], the algorithm based on QR decomposition in Ref. [164] and the algorithm in this chapter meet the requirement of the watermark invisibility, and however the extracted watermark in the former two algorithms is inferior to the algorithm in this chapter. In order to further verify the execution efficiency and the robustness of the proposed algorithm, the chapter will further compare the algorithms proposed in Refs [130, 164] and others.

8.4.3 Test of the Watermark Robustness

In this section, various attacks such as image compression, cropping, adding noise, scaling, filtering, rotation, blurring, and so on are performed on the watermarked images “Lena” and “Avion” and the proposed method is compared with the related methods [130, 164] to estimate the robustness of the proposed method.

Joint Photographic Expert Group (JPEG) is one of the most common document formats in the Internet and digital products. The compression factors of JPEG range from 0 to 100 and when the compression factor is gradually reduced from 100 to 0,the compression result and quality of image are decreased significantly. In this experiment, the watermarked images are compressed with different compression factors from 10 to 100 increasing in steps of 10. Meanwhile, the watermarked images are also performed by JPEG 2000 compression, with the compression ratio from 1 to 10 increasing in steps of 1. Figure 8.6 gives part of the results. Compared with the methods in Refs [130, 164], the proposed scheme achieves a better robustness against the common image compression.

The salt and pepper noise with the intensity of 0.02 and 0.10 is used respectively, to attack the watermarked images. Moreover, the Gaussian noise with the mean value of 0.1 and 0.3 is also added to corrupt the watermarked images. The extracted watermark after being attacked is shown in Figure 8.7.

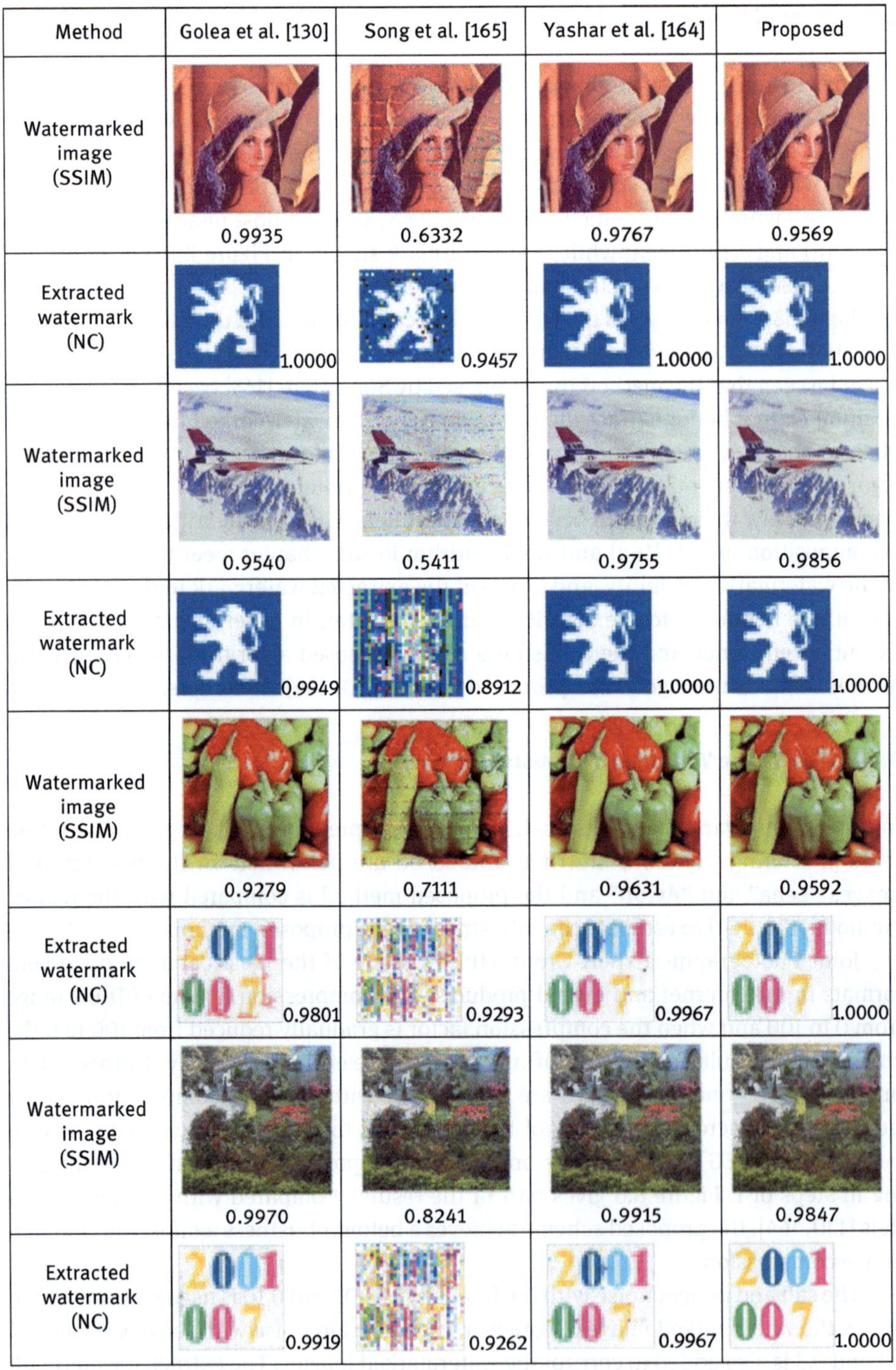

Figure 8.5: The watermarked images (SSIM) and the extracted watermarks via using different methods without any attacks (NC).

Attack	Image	Golea et al. [130]	Yashar et al. [164]	Proposed
JPEG (30)	Lena	 0.6531	 0.8085	 0.9139
	Avion	 0.8204	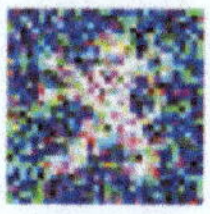 0.8186	 0.8834
JPEG (90)	Lena	 0.9712	 0.9995	 0.9999
	Avion	 0.9760	 0.9982	 0.9999
JPEG 2000 (5:1)	Lena	 0.9390	 0.9949	 0.9995
	Avion	 0.9375	 0.9959	 0.9999
JPEG 2000 (10:1)	Lena	 0.8071	 0.9261	 0.9993
	Avion	 0.7951	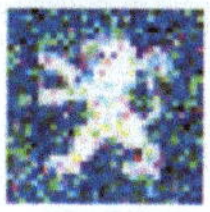 0.9129	 0.9867

Figure 8.6: The extracted watermarks and NC values via using different methods after JPEG compression attack and JPEG 2000 compression attack.

Figure 8.8 shows the experimental results after median filtering and low-pass filtering operation with different parameters. As can be seen from these results, the proposed method has better robustness than other algorithms.

Figure 8.9 not only gives the result of sharpening attacks but also shows the results of blurring attacks. In the procedure of sharpening, the radii are 0.2 and 1.0,

Attack	Image	Golea et al. [130]	Yashar et al. [164]	Proposed
Salt & Peppers noise (0.02)	Lena	 0.5698	 0.8093	 0.9414
	Avion	 0.5276	 0.8089	 0.9226
Salt & Peppers noise (0.10)	Lena	0.2724	 0.5559	 0.7504
	Avion	 0.2345	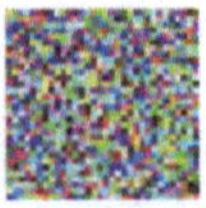 0.5784	 0.7229
Gaussian noise (0.1)	Lena	 0.8600	 0.7084	 0.9817
	Avion	 0.8188	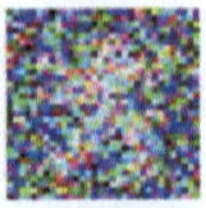 0.7089	 0.9835
Gaussian noise (0.3)	Lena	 0.7469	 0.5492	 0.8598
	Avion	 0.6511	 0.5578	 0.8614

Figure 8.7: The extracted watermarks and NC values via using different methods after adding noise attack.

respectively. In the blurring attacks, two cases are simulated here to degrade the two watermarked images. The radius in the first case is 0.2, while the second case is 1.0. The experimental results show that the proposed method is superior to the method in Refs [130, 164].

Attack	Image	Golea et al. [130]	Yashar et al. [164]	Proposed
Median filter (3×1)	Lena	 0.7102	 0.9919	 0.9993
	Avion	 0.7455	 0.9721	 0.9972
Median filter (5×1)	Lena	 0.5019	 0.9578	 0.9906
	Avion	 0.5168	 0.9118	 0.9765
Low-pass Filter (100,1)	Lena	 0.5477	 0.8901	 0.9676
	Avion	 0.5855	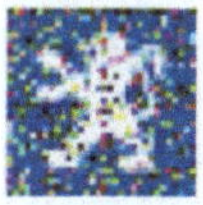 0.8622	 0.9586
Low-pass Filter (100,3)	Lena	 0.3818	 0.8809	 0.8980
	Avion	 0.4650	 0.8431	 0.8686

Figure 8.8: The extracted watermarks and NC values via using different methods after filtering attack.

Figure 8.10 includes the results of scaling and cropping. Two scaling operations of 400 % and 25 % are utilized to deteriorate the watermarked image, respectively. In the cropping attack, because the watermark used by the method in Ref. [130] is not permuted, the position and size of cropping can fully affect the watermark in the cropped area. There is a black area in the extracted watermark, which means the watermark information in this area is fully deleted by the cropping attack.

Attack	Image	Golea et al. [130]	Yashar et al. [164]	Proposed
Sharpening (0.2)	Lena	 0.8481	 0.9999	 0.9999
	Avion	 0.8503	 0.9959	 0.9999
Sharpening (1.0)	Lena	 0.7808	 0.8735	 0.9838
	Avion	 0.6256	 0.8648	 0.9662
Blurring (0.2)	Lena	 1.0000	 0.9912	 1.0000
	Avion	 0.5719	 0.9958	 0.9995
Blurring (1.0)	Lena	 0.2702	 0.7573	 0.7111
	Avion	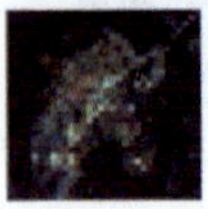 0.1785	 0.7429	 0.6286

Figure 8.9: The extracted watermarks and NC values via using different methods after sharping attack and blurring attack.

In order to test the robustness against rotation attacks, two experiments are investigated in Figure 8.11. One involves rotating the watermarked image clockwise 5°, the other clockwise 30°. During every rotation attack, the watermarked images are first rotated a certain degrees clockwise and then are rotated the estimated degrees

Attack	Image	Golea et al. [130]	Yashar et al. [164]	Proposed
Scaling (4)	Lena	 0.8385	 0.9962	 0.9999
	Avion	 0.8689	 0.9959	 0.9999
Scaling (1/4)	Lena	 0.5698	 0.6124	 0.9838
	Avion	 0.6146	 0.8648	 0.9662
Cropping (25%)	Lena	 0.7380	 0.7586	 0.8772
	Avion	 0.7361	 0.7568	 0.8770
Cropping (50%)	Lena	 0.5311	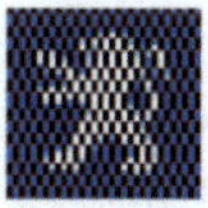 0.5047	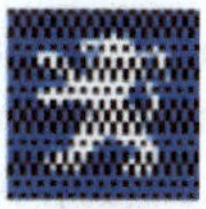 0.6264
	Avion	 0.5331	 0.5024	 0.6264

Figure 8.10: The extracted watermarks and NC values via using different methods after scaling attack and cropping attack.

counterclockwise. After that the cropping and scaling operations are also needed to make the size of the watermarked image as 512 × 512 for watermark extraction. Such rotation attacks simulate the occurrence of truncation errors that degrade the image.

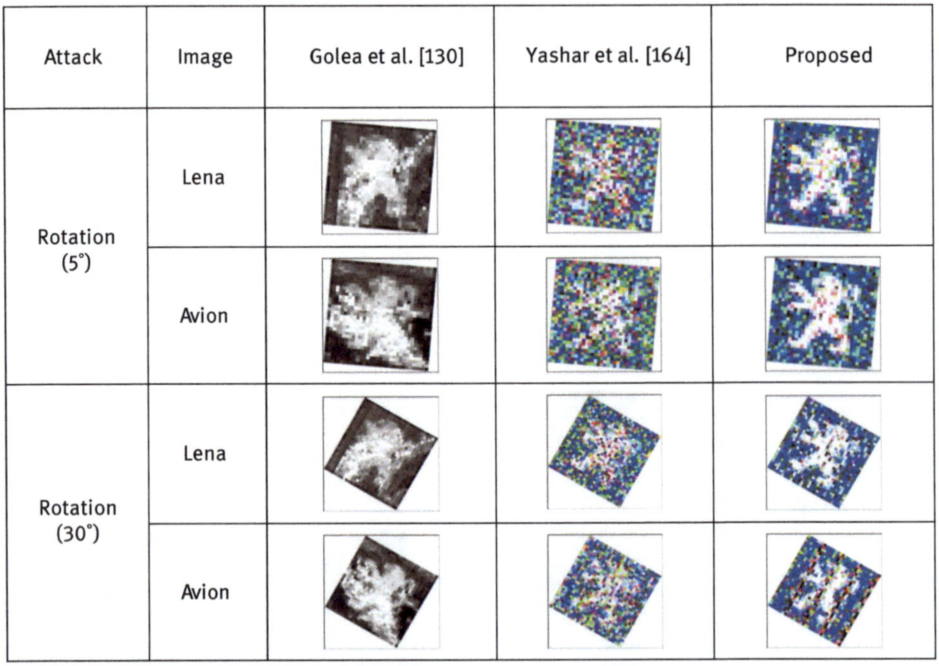

Figure 8.11: The extracted watermarks via using different methods after rotation attack.

8.4.4 Comparison with the Spatial Domain Algorithm

In order to further prove the robustness, the proposed methods are also compared with the spatial domain method [83]. In this spatial domain method, each color pixel's color quantization index is modified to carry the watermark in the embedding process, while the color gamut and quantization table are required in the extraction, which cannot achieve the purpose of blind extraction.

In Ref. [83], two color images of Figure 7.3(c) and 7.3(d) were taken as host image, and 8-color image shown in Figure 7.3(f) was taken as the color watermark image. For fair comparison, we also use the watermark image to carry out the experiment with same attack styles by the method in Ref. [83]. The result in Figure 8.12 reveals that the proposed algorithm has better robustness. This is because the changed color values for various attacks have directly affected the mapping relation between the original color value and the color table, which results in the degraded quality of extracted watermark of the algorithm [83].

8.4.5 The Execution Time Comparison of Different Algorithms

In our experiments, a laptop computer with a duo Intel CPU at 2.27 GHZ, 2.00 GB RAM, Win 7, MATLAB 7.10.0 (R2010a) is used as the computing platform. As shown

Image	Attack	Chou et al. [83]	Proposed method
Peppers	Low-pass filtering	 0.539	 0.9864
	Crop 50%	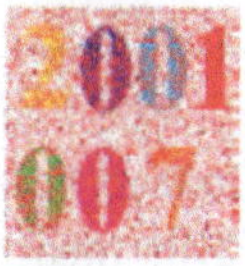 0.553	 0.6238
	Scaling 1/4	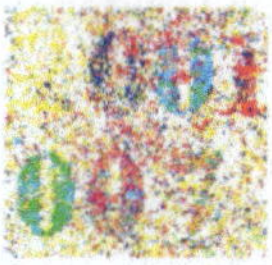 0.536	 0.9659
	Scaling 4	 0.851	 0.9965
	Rotation 30°		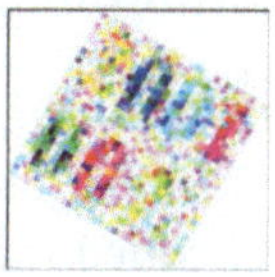
	JPEG 27.5:1	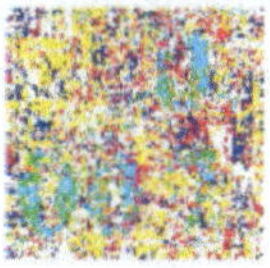 0.343	 0.9963
TTU	Low-pass filtering	 0.423	 0.9651
	Gaussian noise 4	 0.982	 0.9678
	Median Filter 3×3	 0.170	 08039

Figure 8.12: The extracted watermarks via using different methods after different attacks.

Table 8.2: The comparison of performing time between different methods (seconds).

Method	Embedding time	Extraction time	Total time
Scheme [130]	1.909066	0.905951	2.815017
Scheme [83]	1.406568	1.105751	2.512319
Scheme [164]	0.913247	0.455445	1.368692
Proposed method	0.686992	0.427006	1.113998

in Table 8.2, the embedding and extraction time of the proposed method is less than that of the method in Ref. [130]. This means that SVD is more complex than QR decomposition, which is because the QR decomposition is an intermediate step in SVD. Meanwhile, the embedding and extraction time of the proposed algorithm are also less than that of the method in Ref. [83], which is because the host image needs to be transformed to the CIE-Lab color space for the color quantization and its inverse transformation is also required in the method in Ref. [83]. Since the wavelet transform and QR decomposition are involved in the method in Ref. [164], the proposed method using only QR decomposition can cost less time.

8.5 Conclusion

In this chapter, a novel double-color algorithm based on QR decomposition has been proposed. Through using the quantization technique, the color watermark information is embedded into the first row of the fourth column entry in the matrix R by QR decomposition. Moreover, without resorting to the original host image or original watermark, the embedded watermark can be successfully extracted from the images attacked by different attacks. The result of experiment shows that the proposed algorithm not only attains higher invisibility of the watermarking, but consumes less time and has high execution efficiency.

9 The Color Image Watermarking Algorithm Based on Hessenberg Decomposition

It is always a challenging work to research and design a double-color image watermarking algorithm with blind manner, which is different from the majority of existing algorithms that use binary image or gray-scale image as digital watermarks. In this chapter, the feature of the Hessenberg matrix is analyzed and the color image watermarking algorithm based on Hessenberg decomposition is proposed. The encrypted color image watermark information is embedded into the greatest coefficient of the Hessenberg matrix with quantization technique. Moreover, the original host image or the original watermark image is not necessary during the process of watermark extraction. Experimental results show that the proposed watermarking algorithm has a better watermark performance in the aspects of the invisibility, robustness, and computational complexity.

9.1 Introduction

With the rapid development of Internet and multimedia technology, illegal copying and tampering with copyright maliciously have been a serious issue, a technique that is in great need to avoid it. Digital watermarking is considered as an effective method to solve this problem [166]. The essence of digital watermarking is to hide meaningful signals in the host media (such as video, image, audio, and text) to justify the host media copyright information [167].

Recently, the color image watermarking technology has become one of the hotspots in the field of information hiding [39, 43, 56, 168–171]. For example, FindIk et al. [39] proposed to embed the binary image with size of 32 × 32 into the blue components of color image with size of 510 × 510 by using artificial immune recognition system, which has better performance of the watermark. A color image watermarking algorithm [168], based on support vector regression and non-subsampled contourlet transform, was proposed by Niu et al. to resist against geometric attacks. Here, the binary image with size of 32 × 32 was embedded into the green component of color host image and the embedding intensity was determined by the human visual system (HVS). Vahedi et al. [43] proposed a new color image wavelet watermarking method using the principle of bionic optimization to embed the 64 × 64 binary watermark into 512 × 512 color images. Wang et al. [171] proposed to embed the binary image of size 64 × 64 into blind color image of size 256 × 256 in the quaternion Fourier transform domain. Here, the execution time is growth because it needs the operation of least-squares support vector machine regression model. These above-mentioned algorithms all use binary image as watermark. Shao et al. [169] proposed a combination of encryption/watermarking system based on quaternion. Here, color image or gray image of size 64 × 64 was taken as watermark, and color images of size 512 × 512 was taken

DOI 10.1515/9783110487732-009

as the host image in this algorithm. However, this method belongs to non-blind watermarking because the transform coefficients of host image are needed. Chen et al. [40] proposed a new image encryption and watermarking technology to embed some of the gray-level image watermark into the three channels of color image by the addition and subtraction algorithm of neighbor pixel value to achieve blind watermarking. From what has been discussed above, we can see that the embedded watermark is a binary image or gray-level image in most cases when using color image as the host image.

Recently, many scholars have put forward some digital watermarking algorithms based on matrix decomposition [33, 130, 136, 164, 165, 170–175]. For example, Guo and Prasetyo [170] proposed to embed a gray-level watermark image of the same size as the host image into the singular value matrix of the host image transformed by redundant discrete wavelet transform and singular value decomposition (SVD), and this method was a non-blind watermarking scheme because the principal component of the host image was needed when extracting watermark. Lai [33] designed a novel watermarking method based on HVS and SVD, in which the binary watermark was embedded into gray-level image with size of 512×512 by modifying the certain coefficients of the unitary matrix U. This method has better performance in resisting the adding noise, cropping, and median filtering, but is worse in the aspect of resisting the rotation and scaling, and has false-positive detection problem. Although the method in Ref. [130] proposed a blind color image watermarking algorithm, it is needed to modify one or more singular value to maintain the order of singular value, which may reduce the quality of watermark image. Bhatnagar et al. [136] embedded the gray-level watermark with size of 256×256 into the gray-level image with size of 512×512. This method belongs to non-blind watermarking method and has false-positive detection problem. Naderahmadian et al. [172] proposed a gray-scale image watermarking algorithm based on QR decomposition, whose results show that this method has lower computational complexity and better performance of the watermark, but the embedded watermark is binary logo of size 32×32. Based on Hessenberg decomposition theory, the 64×64 gray-level image was embedded in 256×256 gray-level image by the method in Ref. [173], which belongs to the blind watermarking method. Seddik et al. [174] proposed a blind watermarking method using Hessenberg decomposition, in which the host image was gray-level image. Yashar et al. [164] proposed the image was divided into nonoverlapping image blocks with size of 8×8 and each image block was further decomposed to get R matrix by QR decomposition, then one watermark bit was embedded into all entries of the first row of R matrix, finally, the embedded watermark was 88×88 binary image. In the method described in Ref. [165], the 32×32 binary image was embedded into the Q matrix entries of QR decomposition. It is observed that binary image is used as original watermark by the method described in Ref. [164] or [165].

As is well known, the problem that using these color images to protect copyright needs to be considered urgently since the trademarks or logos of many corporations are colored. When the color image watermark of the same size as binary image is embedded into color host image, the information capacity of color image watermark will increase 24 times than that of binary image watermark and 8 times of gray-level image as the same size, which will directly affect the invisibility and robustness of watermark. In our previous works [166, 175], two different watermarking schemes based on QR decomposition were proposed, respectively. Although the proposed algorithm in Ref. [175] has better performance than the algorithm in Ref. [166], these methods have high computational complexity. Theoretically, the time complexity of SVD or Schur decomposition is higher than that of QR decomposition, and Hessenberg decomposition is an intermediate step in QR decomposition. Therefore, Hessenberg decomposition has lower computational complexity than other decomposition methods and can be used to further study on digital watermarking technology.

Based on the above discussion, this chapter proposes a new blind double-color image watermarking algorithm based on Hessenberg decomposition. By further analyzing Hessenberg decomposition, we found that the biggest energy entry of Hessenberg matrix can be quantified to embed watermark when performing Hessenberg decomposition on the 4×4 pixel matrix. The experimental results show that the blind watermark algorithm proposed in this chapter not only has good invisibility and strong robustness, but has shorter execution time.

9.2 Hessenberg Transform on Image Block

Hessenberg decomposition is the orthogonal decomposition of a general matrix A by orthogonal similarity transformations [173, 176]:

$$A = QHQ^T, \tag{9.1}$$

where Q is an orthogonal matrix and H is an upper Hessenberg matrix, which means $h_{i,j} = 0$ whenever $i > j + 1$. Hessenberg decomposition is usually computed by Householder matrices. The Householder matrix (P) is an orthogonal matrix as follows:

$$P = (I_n - 2uu^T)/u^T u, \tag{9.2}$$

where u is a nonzero vector in R^n and I_n is the $n \times n$ identity matrix. There are n–2 steps in the overall procedure when matrix A is of size $n \times n$. Therefore, Hessenberg decomposition is represented as follows:

$$H = (P_1P_2 \cdots P_{n-3}P_{n-2})^T A(P_1P_2 \cdots P_{n-3}P_{n-2}), \tag{9.3}$$

$$\Rightarrow H = Q^T AQ, \tag{9.4}$$

$$\Rightarrow A = QHQ^T, \tag{9.5}$$

where $Q = P_1P_2 \cdots P_{n-3}P_{n-2}$.

For example, the following matrix A with size of 4×4 is an original pixel block:

$$A = \begin{bmatrix} 80 & 91 & 91 & 95 \\ 83 & 89 & 88 & 96 \\ 90 & 89 & 89 & 96 \\ 96 & 93 & 88 & 95 \end{bmatrix}. \tag{9.6}$$

When A is decomposed by Hessenberg transform, its orthogonal matrix Q, and upper Hessenberg matrix H are given as follows:

$$Q = \begin{bmatrix} 1 & 0 & 0 & 0 \\ 0 & -0.5335 & 0.7622 & 0.3667 \\ 0 & -0.5785 & -0.0125 & -0.8156 \\ 0 & -0.6170 & -0.6473 & 0.4476 \end{bmatrix}, \tag{9.7}$$

$$H = \begin{bmatrix} 80.0000 & -159.8089 & 6.7321 & 1.6707 \\ -155.5796 & 273.8047 & -10.2233 & -6.7820 \\ 0 & -15.1564 & -1.9211 & -0.2571 \\ 0 & 0 & 1.6583 & 1.1164 \end{bmatrix}. \tag{9.8}$$

In the above Hessenberg matrix H, the coefficient 273.8047 with the biggest energy can be suitably modified to embed watermark information.

9.3 Proposed Method

The proposed method mainly includes two processes, that is, watermark embedding and watermark extraction, which will be introduced as follows.

9.3.1 Watermark Embedding

Figure 9.1 illustrates the embedded watermarking scheme based on Hessenberg decomposition, and the detailed embedding steps are listed as follows.

1. Converting the color image watermark to binary information
   First, the original color watermark image $W$with size $p \times q$ is partitioned to three two-dimensional watermark components of R, G, and B by dimensionality reduction; second, each two-dimensional watermark component is scrambled by Arnold transform with private key KA_i ($i = 1, 2, 3$) for improving the security of the watermarking [158]; third, each pixel value is converted to 8-bit binary sequence; finally, all 8-bit binary sequences are combined to obtain the binary sequence W_i ($i = 1, 2, 3$).

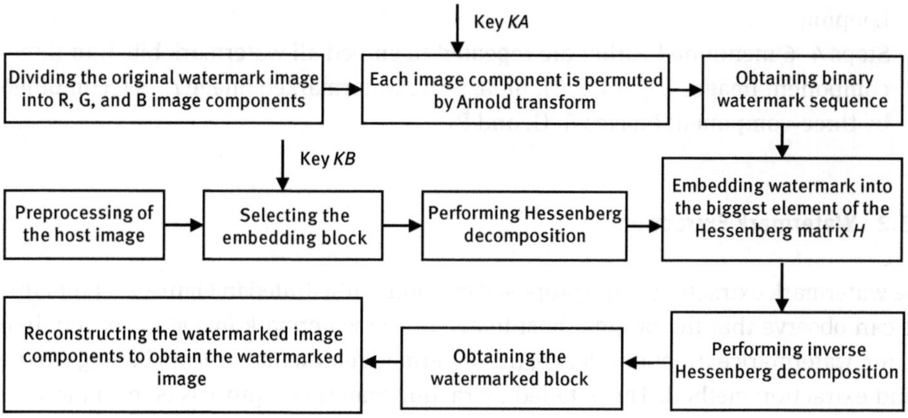

Figure 9.1: The diagram of watermark embedding.

2. Preprocessing the host image
   When embedding watermark, the host image I is also partitioned to three-component images of R, G, and B, denoted as I_j $(j = 1, 2, 3)$, and each component image I_j is further divided into nonoverlapping image blocks with size of 4×4.
3. Selecting the random blocks for embedding watermark
   For improving the robustness and security of the method, Hash pseudorandom replacement algorithm is, which is based on MD5 with private key KB_i $(i = 1, 2, 3)$, used to randomly select the embedding image blocks [128].
4. Performing Hessenberg decomposition
   According to eq. (9.1), each embedding block is executed by Hessenberg decomposition and its Hessenberg matrix H is obtained.
5. Embedding watermark
   Search for the biggest energy entry $h_{\max}$ of the Hessenberg matrix H, then use eq. (9.9), and the watermark w is embedded into the selected block by modifying $h_{\max}$ with $h^*_{\max}$:

$$h^*_{\max} = \begin{cases} h_{\max} - \mathrm{mod}(h_{\max}, T) + 0.75 * T, & \text{if } w = "1" \\ h_{\max} - \mathrm{mod}(h_{\max}, T) + 0.25 * T, & \text{if } w = "0" \end{cases}, \tag{9.9}$$

   where mod(.) is the module operation function and T is the threshold.
6. Performing inverse Hessenberg operation
   Use eq. (9.10) to inverse Hessenberg operation and obtain the watermarked image block A^*:

$$A^* = QH^*Q^T. \tag{9.10}$$

7. Looping
   Steps 4–6 mentioned earlier are repeated to embed all watermark bits into three-component images I_j ($j = 1, 2, 3$). At last, the watermarked image I^* is recombined by three-component images R, G, and B.

9.3.2 Watermark Extraction

The watermark extraction of the proposed method is illustrated in Figure 9.2. From this we can observe that the original host image or the watermark image is not required during watermark extraction. Hence, the watermark extraction process belongs to the blind extraction method. The detailed extraction watermark process is introduced as follows:

1. Preprocessing the watermarked image
   At first, the watermarked image I^* is partitioned to the three-component images of R, G, and B. Then each component image is further divided into nonoverlapping image blocks with size of 4×4.
2. Selecting the watermarked image block
   The MD5-based Hash pseudorandom replacement algorithm with private key KB_i ($i = 1, 2, 3$) is used to select the watermarked image blocks.
3. Performing inverse Hessenberg operation
   According to eq. (9.1), each embedded block is decomposed by Hessenberg transform and its upper Hessenberg matrix H is obtained.
4. Extracting watermark
   Search for the biggest energy entry h^*_{max} of the Hessenberg matrix H^*, then use the following equation to extract the watermark information w^*:

$$w^* = \begin{cases} \text{"0"}, & \text{if } \bmod(h^*_{max}, T) < 0.5 * T \\ \text{"1"}, & \text{else} \end{cases}. \tag{9.11}$$

   Steps 2–4 mentioned earlier are repeated until extracting all watermarks.

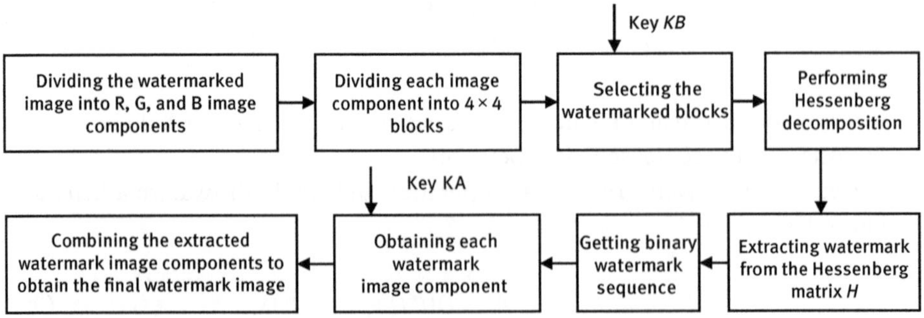

Figure 9.2: The diagram of watermark extraction.

5. Obtaining the component watermark
   The extracted information w^* is divided into some binary sequence size of 8 bits, and each binary sequence is converted to decimal number, then the component watermark w^* $(j = 1, 2, 3)$ is formed.
6. Forming the final watermark
   Each component of the watermarking is transformed by the anti-Arnold transformation with the private key KA_i $(i = 1, 2, 3)$ and combined into the final extracted watermark w^*.

9.4 Algorithm Testing and Result Analysis

The performance of the proposed method is measured by the invisibility, robustness, capacity, and computational complexity. To validate the watermark performances of the proposed algorithm, all 24-bit 512 × 512 color images of the image database CVG-UGR are used as the host images, and two 24-bit 32 × 32 color images shown in Figure 9.3 are used as original watermarks. For relatively fair comparison, the host images in Figure 9.4 used in Refs [130, 175] are adopted in this experiment to make a comparison. All the schemes are coded in MATLAB 2010a and performed on a laptop computer with Intel Core CPU @ 2.27 GHZ, 2.00 GB RAM.

In order to evaluate the watermark invisibility, this chapter not only uses the traditional measure peak signal-to-noise ratio (PSNR), but also takes the structural similarity index measurement (SSIM) as a new way to measure the similarity between the original color image I and the watermark image I^*.

In this algorithm, the selection of quantization step plays an important role. For deciding the quantization step T, the selected standard images are used in the experiment. As shown in Figure 9.5, with the increase of the quantization step, the

(a) (b)

Figure 9.3: Watermark images: (a) Peugeot logo and (b) 8-color image.

(a) (b) (c) (d)

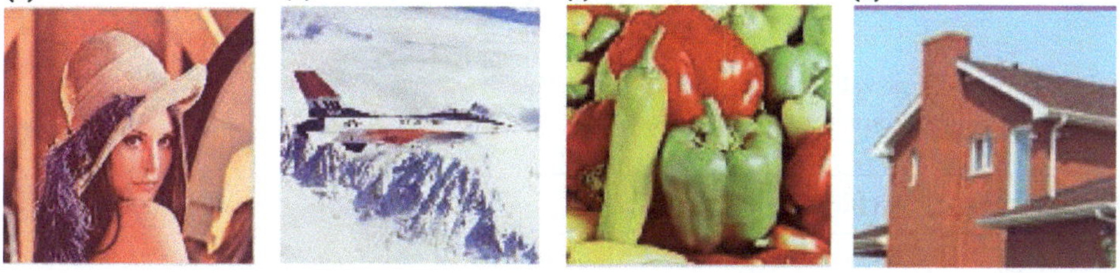

Figure 9.4: Original host images: (a) Lena, (b) Avion, (c) Peppers, and (d) House.

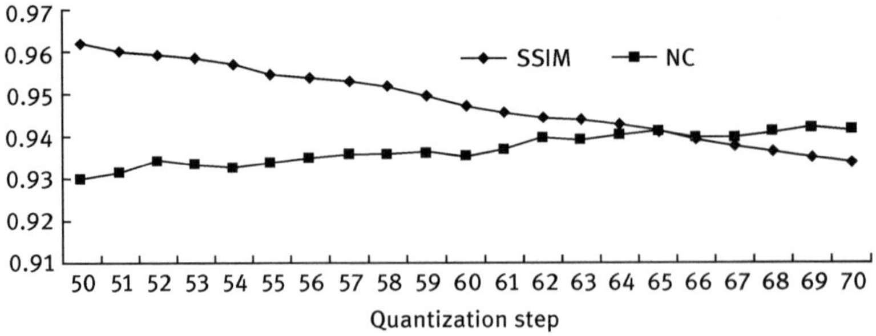

Figure 9.5: The results of invisibility and robustness with different quantization steps.

SSIM value is decreasing but the normalized cross-correlation (NC) value is increasing, which means the watermark imperceptibility is weaker and weaker but the watermark robustness gets stronger. Considering the trade-off between the invisibility and the robustness of watermark, the quantization step is set to 65. It is noted that the NC value in Figure 9.5 is the average NC of the extracted watermark from all attacked watermark images.

9.4.1 Test of the Watermark Invisibility

In general, a watermarked image with larger PSNR and SSIM shows that watermarked image is more similar to the original host image, which means that the watermarking algorithm has good watermark invisibility. A larger NC shows that the extracted watermark is more similar to the original watermark, and the algorithm is more robust. For the proposed algorithm in this chapter, the above-mentioned color image watermarks are embedded into the host images with size of 512 × 512 from image database CVG-UGR with quantization step $T = 65$. Table 9.1 not only lists the average PSNR and SSIM values of all watermarked images, but also gives the average NC values of extracted watermark images, which shows that the watermark image can be hidden to the host image well.

For further verifying the watermark invisibility of the proposed algorithm, we will compare with the methods in Refs [130, 164, 165, 175] using different host images and watermark images. From Figure 9.6, we can see that the watermarking algorithm based on QR decomposition proposed in Ref. [165] cannot extract the original watermark and

Table 9.1: The average values in terms of PSNR, SSIM, and NC after using different watermark images.

Watermark	PSNR (dB)	SSIM	NC
Figure 9.3(a)	37.3922	0.9426	0.9984
Figure 9.3(b)	37.5258	0.9438	0.9979

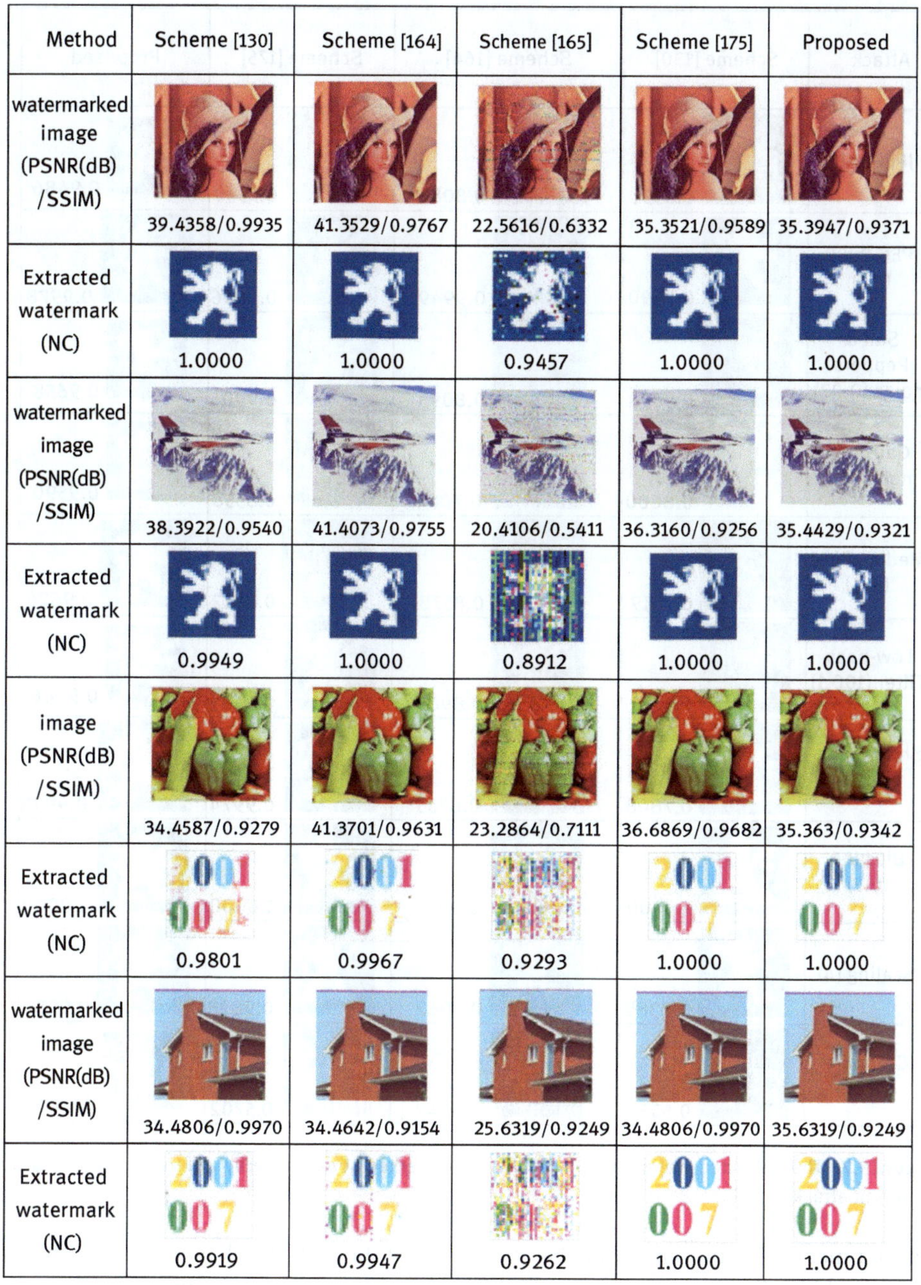

Method	Scheme [130]	Scheme [164]	Scheme [165]	Scheme [175]	Proposed
watermarked image (PSNR(dB) /SSIM)	39.4358/0.9935	41.3529/0.9767	22.5616/0.6332	35.3521/0.9589	35.3947/0.9371
Extracted watermark (NC)	1.0000	1.0000	0.9457	1.0000	1.0000
watermarked image (PSNR(dB) /SSIM)	38.3922/0.9540	41.4073/0.9755	20.4106/0.5411	36.3160/0.9256	35.4429/0.9321
Extracted watermark (NC)	0.9949	1.0000	0.8912	1.0000	1.0000
image (PSNR(dB) /SSIM)	34.4587/0.9279	41.3701/0.9631	23.2864/0.7111	36.6869/0.9682	35.363/0.9342
Extracted watermark (NC)	0.9801	0.9967	0.9293	1.0000	1.0000
watermarked image (PSNR(dB) /SSIM)	34.4806/0.9970	34.4642/0.9154	25.6319/0.9249	34.4806/0.9970	35.6319/0.9249
Extracted watermark (NC)	0.9919	0.9947	0.9262	1.0000	1.0000

Figure 9.6: The comparison of the watermarked images and the extracted watermarks via using different methods without any attacks.

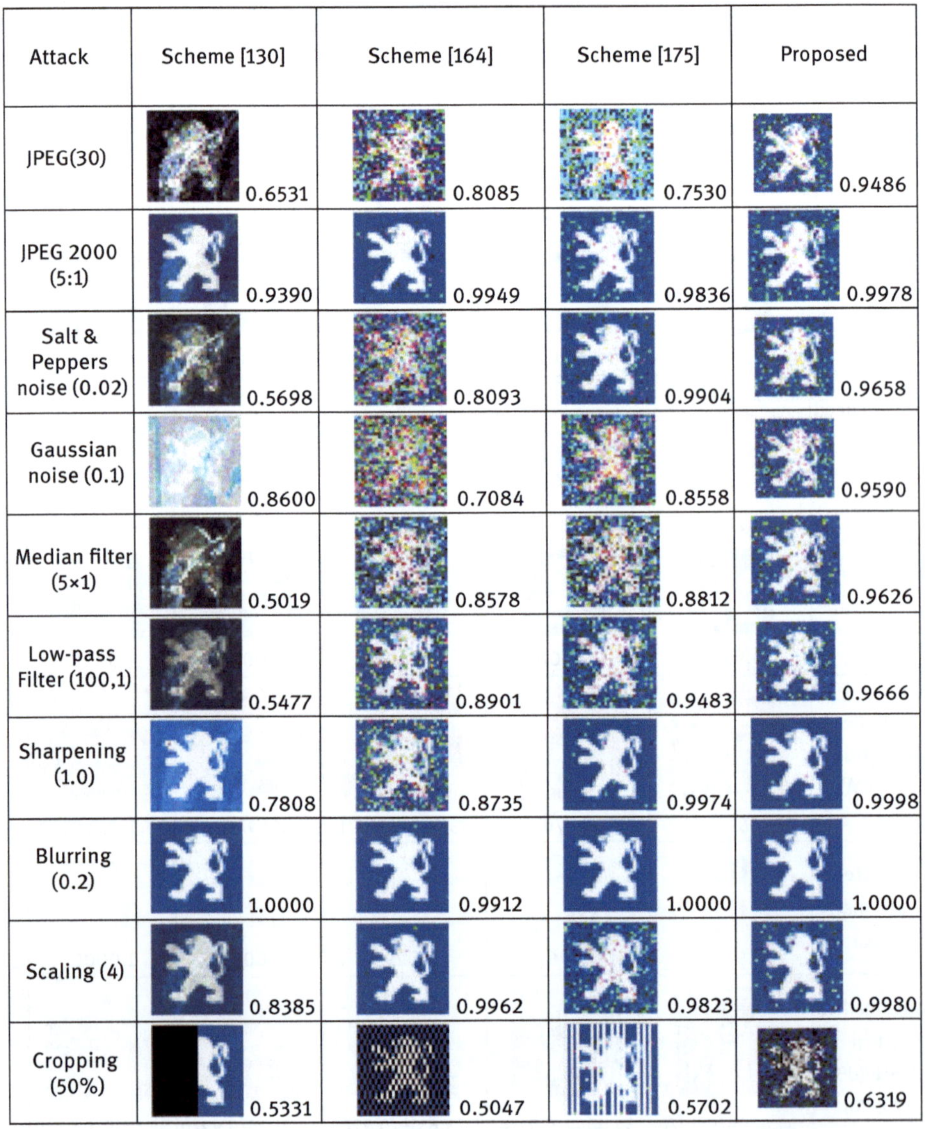

Attack	Scheme [130]	Scheme [164]	Scheme [175]	Proposed
JPEG(30)	0.6531	0.8085	0.7530	0.9486
JPEG 2000 (5:1)	0.9390	0.9949	0.9836	0.9978
Salt & Peppers noise (0.02)	0.5698	0.8093	0.9904	0.9658
Gaussian noise (0.1)	0.8600	0.7084	0.8558	0.9590
Median filter (5×1)	0.5019	0.8578	0.8812	0.9626
Low-pass Filter (100,1)	0.5477	0.8901	0.9483	0.9666
Sharpening (1.0)	0.7808	0.8735	0.9974	0.9998
Blurring (0.2)	1.0000	0.9912	1.0000	1.0000
Scaling (4)	0.8385	0.9962	0.9823	0.9980
Cropping (50%)	0.5331	0.5047	0.5702	0.6319

Figure 9.7: The extracted watermarks from the watermarked Lena image by different methods after different attacks.

cannot meet the requirement of the invisibility of the watermark, so the algorithm is not suitable to embed color image watermark into the color host image. Relatively, other algorithms have better invisibility. Obviously, the proposed algorithm not only meets the requirements of watermark invisibility, but also can effectively extract the embedded watermark. In order to further prove the robustness of the proposed algorithm, we have to compare it with the methods described in Refs [130, 164, 175] in the following sections.

9.4.2 Test of the Watermark Robustness

In order to further verify the robustness of the algorithm in this chapter, a variety of attacks are used (such as image compression, shear, adding noise, scaling, filtering, and fuzzy) to attack the watermarked image, and the algorithm is compared with other

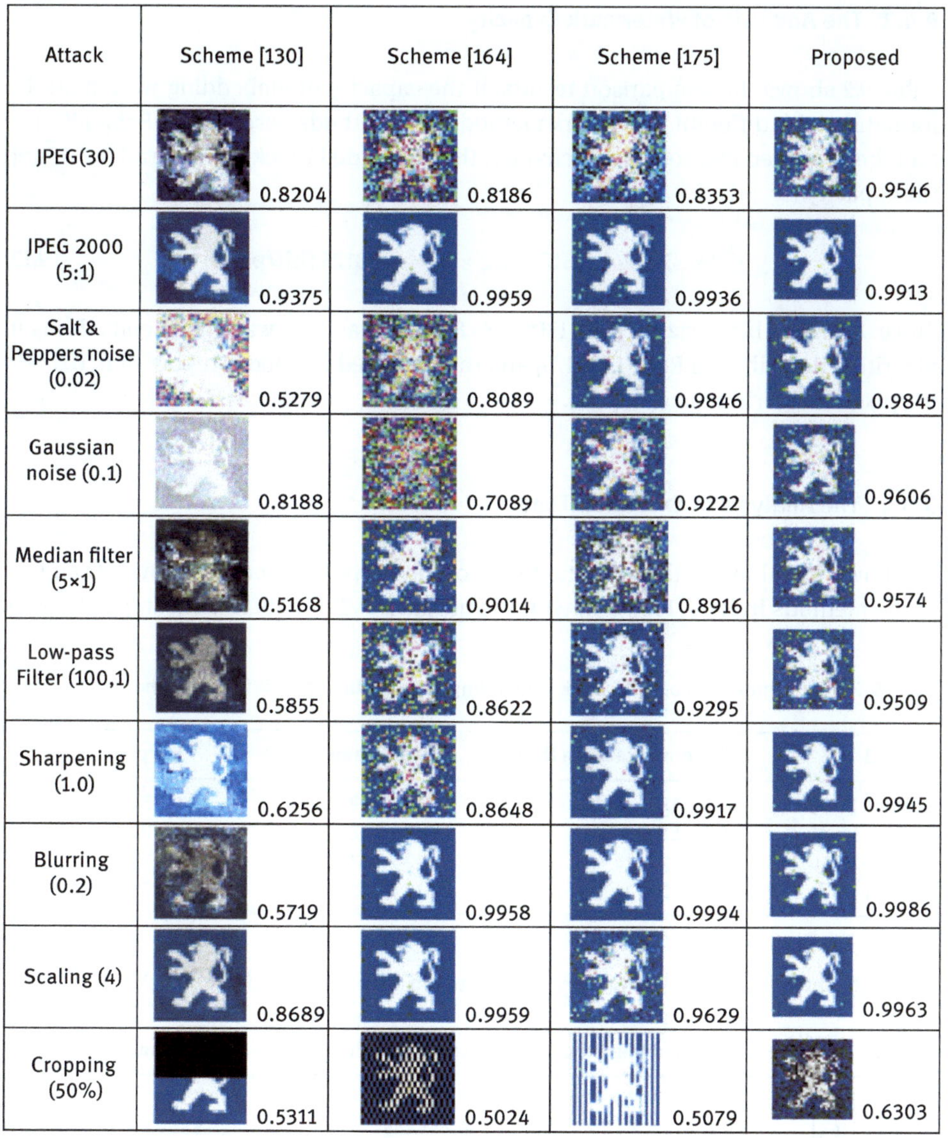

Figure 9.8: The extracted watermarks from the watermarked Avion image by different methods after different attacks.

related algorithms [130, 164, 175]. For saving some space, Figures 9.7 and 9.8 show the result of watermark shown in Figure 9.3(a) extracted from the attacked images "Lena" and "Avion", respectively. The experimental results show that the algorithm is better than other algorithms [130, 164, 175] under most attack tests in the aspect of the robustness.

9.4.3 The Analysis of Watermark Capacity

Table 9.2 shows the comparison results of the capacity of embedding watermark information with different watermark methods. For methods described in Refs [130, 175] and the proposed method in this chapter, the embedded block sizes are all 4×4, and the capacity is

$$(32 \times 32 \times 24) / (512 \times 512 \times 3) = 0.03125 \text{ (bit/pixel)} . \tag{9.12}$$

Since the image block size in Ref. [130] is 8×8, the capacity of watermark embedding in algorithm described in Refs [130, 175] and the proposed method is higher than 0.02954 bit/pixel of algorithm in Ref. [164].

9.4.4 The Analysis of Execution Time

As shown in Table 9.3, the execution time of the proposed method is lower than that of the methods in Refs [130, 175], which is because SVD and Schur decomposition are

Table 9.2: The comparison of watermark embedding capacity between different methods.

Method	Watermark length (bit)	Host image (pixel)	bit/pixel
Scheme [130]	32 × 32 × 24	512 × 512 × 3	0.03125
Scheme [164]	88 × 88	512 × 512	0.02954
Scheme [175]	32 × 32 × 24	512 × 512 × 3	0.03125
Proposed	32 × 32 × 24	512 × 512 × 3	0.03125

Table 9.3: The comparison of performance time between different methods (seconds).

Method	Embedding time	Extraction time	Total time
Scheme [130]	1.909066	0.905951	2.815017
Scheme [164]	0.913247	0.455445	1.368692
Scheme [175]	0.686992	0.427006	1.113998
Proposed	0.566331	0.320511	0.886842

more complex than QR decomposition, and Hessenberg decomposition is an intermediate step of QR decomposition. At the same time, since the method in Ref. [164] used the wavelet transform and QR decomposition, the proposed algorithm that used only Hessenberg decomposition can spend less time.

9.5 Conclusion

In this chapter, we have proposed a new algorithm based on Hessenberg decomposition for embedding the color image watermark into the color host image. On the basis of the image pixel blocks with Hessenberg decomposition, color watermark messages are embedded in the biggest entry of Hessenberg matrix H by modifying the maximum energy entries. In addition, the embedded watermark can be extracted from the different attacked images successfully without resorting the original host image or original watermark. Experimental results have shown that the proposed color image watermarking algorithm not only has reached the requirement of watermark invisibility, but also has strong robustness for common image processing and geometric attacks.

10 Summary and Prospect

10.1 Summary

Color image digital watermarking technology not only has a strong application value and developing prospects but also its methods and technologies in the research process provide an effective reference for other related technology. At present, this technology has become one of the hotspots in the field of information security, although many researchers at home and abroad in this field carried out effective work and achieved many meaningful research and application results, because the digital color image itself contains a very large amount of information and most of the current algorithms focus on non-blind extraction, which makes this technology to become one of the difficulties in the field of digital watermarking. How to improve the robustness and efficiency of the watermarking algorithm is a key problem in the premise of the color watermark that can be invisibly embedded.

The book summarizes the current state of color image watermarking algorithm; starts a more in-depth research on the watermarking algorithm from the aspects of the watermark embedding capacity, invisibility, robustness, and time complexity; makes some innovative research results; and has theoretical and practical significance in promoting the use of digital watermarking in copyright protection based on analyzing the existing research work of the existent problems and shortcomings, and using combining domain technology, state encoding technology, and matrix decomposition technology based on image block. Of course, the research work of this book still has many drawbacks, which needs further improvement and supplement.

The main research work of this book is summarized as follows:

1. A new color image watermarking algorithm based on discrete cosine transform (DCT) in spatial domain has been proposed in this book. According to the generating principle and the distribution feature of direct current (DC) coefficient in DCT transformation, DC coefficients of each 8×8 block are directly calculated in spatial domain without DCT, and each watermark information bit is embedded repeatedly 4 times by means of the coefficient quantization method. After extracting the watermark, the final binary watermark is determined by the principles of "first to select then to combine" and "the minority obeying the majority." In this method, the watermark can be extracted from the watermarked image without the requirement of the original watermark or the original host image, which not only has the advantages of high efficiency in spatial domain algorithm, but also has the advantage of strong robustness of transform domain algorithm.
2. A new dual-color watermarking algorithm based on integer wavelet transform (IWT) and state coding has been proposed in this book. The algorithm not only uses the feature that the IWT does not have rounding error, but also uses state coding technology to represent the characteristics of watermark pixel information

DOI 10.1515/9783110487732-010

in the form of nonbinary information. By changing the state code of data set to embed the watermark, the watermark can be directly extracted by using the state code of the data set. The simulation results show that the algorithm can embed high-capacity color watermark information into color host image.

3. A double-color image watermarking method based on optimal compensation of singular value decomposition (SVD) has been proposed in this book. The features of SVD are analyzed systematically, and a new optimal compensation of matrix is proposed. When embedding watermark, each 4×4 pixel block is decomposed by SVD, and the watermark bit is embedded into U component by modifying the second row of the first column and the third row of the first column elements of U component. Then the embedded block is compensated by the proposed optimization operation to further improve the invisibility of watermark. When extracting the watermark, the embedded watermark is directly extracted from various attacked images by using the relation between the modified elements of U component without resorting to the original data. Experimental results show that the proposed watermarking algorithm not only overcomes the shortcoming of false-positive detection in other SVD watermarking methods, but also has strong robustness.
4. A color image blind watermarking algorithm based on Schur decomposition has been proposed in this book. First, the theory of Schur decomposition and the feature of decomposed image block are further studied. Then the relationship between the coefficients is not only modified to embed watermark, but also used to extract blind watermark. Experimental results show that not only the invisibility of the proposed algorithm is ensured, but also the robustness is significantly enhanced.
5. An efficient blind watermarking algorithm for color images based on QR decomposition has been proposed in this book. First, the color host image is divided to 4×4 nonoverlapping pixel blocks. Then each selected pixel block is decomposed by QR decomposition and the first row of the fourth column element in the matrix R is quantified for embedding the watermark information. In the process when extracting watermark, the watermark can be extracted from the watermarked image without the original host image or the original watermark image. The simulation results show that the algorithm not only meets the requirements of the invisibility and robustness of the watermark performance, but also has high efficiency in computation complexity.
6. A very invisible color image watermarking algorithm based on Hessenberg decomposition has been proposed in this book. The main principle is to decompose the selected 4×4 nonoverlapping pixel blocks by Hessenberg decomposition and embed the watermarking information into the biggest coefficient of Hessenberg matrix by quantization technology. When extracting the watermark, the original host image or the original watermark image is not needed. The experimental results show that the proposed watermarking algorithm has the obvious advantages

in the invisibility of watermark; other performances can also meet the demand of the watermarking algorithm.

10.2 Prospect

Because the research on digital watermarking technology of color image involves many subjects, and the involved theories are more difficult, there are some imperfections in the research. Hence, there are still a lot of work to be done in the practical application and further research:

1. Improving the performance to resist geometric attacks. At present, although the proposed color image watermarking algorithms with blind manner have strong robustness against the processing of common image and have good ability to resist some geometric attacks (such as cropping and resizing), further improvement in the algorithm performance against rotation attack with large rotation angle is needed.
2. Improving the practicality of the algorithm. At present, the proposed color image watermarking algorithms with blind manner is designed and implemented on the simulation environment; and it is needed to further optimize the performance of the algorithm for reducing the computation complexity and improving the real-time performance, and to make the realization algorithm in software and hardware system.
3. Extending the research object about blind extraction. The proposed watermarking blind extraction algorithms mainly use the color image as the research object in this book. It is necessary to further research on how to use the watermarking algorithm to realize the blind watermarking extraction on digital video, audio, dynamic image, and other multimedia information efficiently.

The blind watermarking technology of color image involves different disciplines and theories, has strong practical value, and brings a lot of new challenges to researchers and engineering technicians. Although some preliminary results are proposed in this book, the proposed algorithms are also not perfect and mature; at the same time, some negligence and deficiencies are also present in the book due to the limit of time and our theory of knowledge, and I sincerely hope all experts and scholars can give me criticism. I would like to express my sincere gratitude again!

References

[1] Wang B, Cheng Q, Deng F. Digital watermarking technique. Xi'an: The Press of Xi'an University of Electronic Science and Technology, 2003.

[2] Sun S, Lu Z, Niu X. Digital watermarking technique and applications. Beijing: Science Press, 2004.

[3] Cox I J, Miller M L, Bloom J A. Digital watermarking. Translated by Wang Y, Huang Z. Beijing: Publishing House of Electronics Industry, 2003.

[4] Lu C S. Multimedia security: Steganography and digital watermarking techniques for protection of intellectual property. Northern California: Idea Group Publishing, 2005.

[5] Chen M, Niu X, Yang Y. The research developments and applications of digital watermarking. Journal of China Institute of Communications, 2001, 22, 71–79.

[6] Jin X Z. Research on digital watermarking algorithm based on DCT domain. Jilin: Jilin University, 2011.

[7] Shikata J, Matsumoto T. Unconditionally secure steganography against active attacks. IEEE Transactions on Information Theory, 2008, 54, 2690–2705.

[8] Wang D, Liang J, Dai Y, Luo S, Qi D. Evaluation of the validity of image watermarking. Chinese Journal of Computers, 2003, 26, 779–788.

[9] Rastegar S, Namazi F, Yaghmaie K, Aliabadian A. Hybrid watermarking algorithm based on Singular Value Decomposition and Radon transform. AEU-International Journal of Electronics and Communications, 2011, 65, 658–663.

[10] Holliman M, Memon N. Counterfeiting attacks on oblivious block-wise independent invisible watermarking schemes. IEEE Transactions on Image Processing, 2000, 9, 432–441.

[11] Lin C Y, Chang S F. Watermarking capacity of digital images based on domain-specific masking effects. International Conference on Information Technology: Coding and Computing, April 2–4, 2001, Las Vegas, NV, 90–94.

[12] Zeng W, Liu B. A statistical watermark detection technique without using original images for resolving rightful ownerships of digital images. IEEE Transactions on Image Processing, 1999, 8, 1534–1548.

[13] Wong P W. A public key watermark for image verification and authentication. International Conference on Image Processing, October 4–7, 1998, Chicago, 455–459.

[14] Fridrich J, Goljan M. Images with self-correcting capabilities. International Conference on Image Processing, October 24–28, 1999, Kobe, 792–796.

[15] Yin P, Yu H H. A semi-fragile watermarking system for MPEG video authentication. IEEE International Conference on Acoustics, Speech, and Signal Processing, May 13–17, 2002, Orlando, FL, IV 3461–3464.

[16] Sun Q, Chang S F, Maeno K, Suto M. A new semi-fragile image authentication framework combining ECC and PKI infrastructures. IEEE International Symposium on Circuits and Systems, May 26–29, 2002, Phoenix-Scottsdale, AZ, 440–443.

[17] Jafri S A R, Baqai S. Robust digital watermarking for wavelet-based compression. The 9th IEEE Workshop on Multimedia Signal Processing, October 1–3, 2007, Crete, 377–380.

[18] Campisi P, Kundur D, Neri A. Robust digital watermarking in the ridgelet domain. IEEE Signal Processing Letters, 2004, 11, 826–830.

[19] Braudaway G W, Magerlein K A, Mintzer F C. Protecting publicly available images with a visible image watermark. Electronic Imaging, Science & Technology. International Society for Optics and Photonics, 1996, 126–133.

[20] Hu Y, Jeon B. Reversible visible watermarking and lossless recovery of original images. IEEE Transactions on Circuits and Systems for Video Technology, 2006, 16, 1423-1429.

[21] Hsu C T, Wu J L. Hidden digital watermarks in images. IEEE Transactions on Image Processing, 1999, 8, 58–68.

[22] Barni M, Bartolini F, Rosa A D, Piva A. Optimum decoding and detection of multiplicative watermarks. IEEE Transactions on Signal Processing, 2003, 51, 1118–1123.

[23] Serdean C V, Ambroze M A, Tomlinson M, Wade J G. DWT-based high-capacity blind video watermarking, invariant to geometrical attacks. IEE Proceedings Vision, Image and Signal Processing, 2003, 150, 51–58.

[24] Huang H C, Wang F H, Pan J S. Efficient and robust watermarking algorithm with vector quantisation. IEEE Electronics Letters, 2001, 37 (13), 826–828.

[25] Zeng G, Qiu Z, Zhang C. Performance analysis of distortion-compensated QIM watermarking. Journal of Electronics & Information Technology, 2010, 32, 86–91.

[26] Yeo I K, Kim H J. Modified patchwork algorithm: A novel audio watermarking scheme. IEEE Transactions on Speech and Audio Processing, 2003, 11, 381–386.

[27] Hartung F, Ramme F. Digital rights management and watermarking of multimedia content for commerce applications. IEEE Communications Magazine, 2000, 38, 80–84.

[28] Lemma A N, Aprea J, Oomen W, Van de Kerkhof L. A temporal domain audio watermarking technique. IEEE Transactions on Signal Processing, 2003, 51, 1088–1097.

[29] Xiang Y, Natgunanathan I, Peng D, Zhou W, Yu S. A dual-channel time-spread echo method for audio watermarking. IEEE Transactions on Information Forensics and Security, 2012, 7, 383–392.

[30] Giakoumaki A, Pavlopoulos S, Koutsouris D. Multiple image watermarking applied to health information management. IEEE Transactions on Information Technology in Biomedicine, 2006, 10, 722–732.

[31] Akleylek S, Nuriyev U. Steganography and new implementation of steganography. Proceedings of the 13th IEEE Signal Processing and Communications Applications Conference, May 16–18, 2005, Ankara, 64–67.

[32] Wang L, Guo C, Li P. Information hiding technique. Hubei: Wuhan University Press, 2004. (China)

[33] Lai C C. An improved SVD-based watermarking scheme using human visual characteristics. Optics Communications, 2011, 284, 938–944.

[34] Wang Z, Bovik A C, Sheikh H R, Simoncelli E P. Image quality assessment: From error visibility to structural similarity. IEEE Transactions on Image Processing, 2004, 13, 600–612.

[35] I. Recommendation 500-11. Methodology for the subjective assessment of the quality of television pictures. International Telecommunication Union, Geneva, Switzerland, 2002.

[36] Muselet D, Tremeau A. Recent trends in color image watermarking. Journal of Imaging Science and Technology, 2009, 53, 0102011–01020115.

[37] Wang X, Lin T, Xue Q. A novel colour image encryption algorithm based on chaos. Signal Processing, 2012, 92, 1101–1108.

[38] Bhatnagar G, Wu Q M J, Raman B. Robust gray-scale logo watermarking in wavelet domain. Computers and Electrical Engineering, 2012, 38, 1164–1176.

[39] Findlk O, Babaoglu I, Ülker E. A color image watermarking scheme based on artificial immune recognition system. Expert Systems with Applications, 2011, 38, 1942–1946.

[40] Chen L, Zhao D, Ge F. Gray images embedded in a color image and encrypted with FRFT and Region Shift Encoding methods. Optics Communications, 2010, 283, 2043–2049.

[41] Wang W, Zuo W, Yan X. New gray-scale watermarking algorithm of color images based on Quaternion Fourier Transform. The 3rd International Workshop on Advanced Computational Intelligence, August 25–27, 2010, Suzhou, China, 593–596.

[42] Rawat S, Raman B. A new robust watermarking scheme for color images. The 2nd IEEE International Advance Computing Conference, February 19–20, 2010, Patiala, 206–209.

[43] Vahedi E, Zoroofi R A, Shiva M. Toward a new wavelet-based watermarking approach for color images using bio-inspired optimization principles. Digital Signal Processing, 2012, 22, 153–162.

[44] Kwitt R, Meerwald P, Uhl A. Color image watermarking using multivariate power exponential distribution. The 16th IEEE International Conference on Image Processing, November 7–10, 2009, Cairo, 4245–4248.

[45] Nasir I, Weng Y, Jiang J. Novel multiple spatial watermarking technique in color images. The 5th International Conference on Information Technology: New Generations, April 7–9, 2008, Las Vegas, NV, 777–782.

[46] Tsui T K, Zhang X P, Androutsos D. Color image watermarking using multidimensional Fourier transforms. IEEE Transactions on Information Forensics and Security, 2008, 3, 16–28.

[47] Liao S. Dual color images watermarking algorithm based on symmetric balanced multiwavelet. International Symposium on Intelligent Information Technology Application Workshops, December 21–22, 2008, Shanghai, China, 439–442.

[48] Luo X Y, Wang D S, Wang P, Liu F L. A review on blind detection for image steganography. Signal Processing, 2008, 88, 2138–2157.

[49] Yong Z, Li C L, Shen L Q, Tao J Z. A blind watermarking algorithm based on block DCT for dual color images. The 2nd International Symposium on Electronic Commerce and Security, December 24–26, 2009, Harbin, China, 213–217.

[50] Tsaia H H, Jhuanga Y J, Lai Y S. An SVD-based image watermarking in wavelet domain using SVR and PSO. Applied Soft Computing, 2012, 12(8), 2442–2453.

[51] Mao J, Lin J, Dai M. An attacked image based hidden messages blind detect technique. Chinese Journal of Computers, 2009, 32, 318–327.

[52] Chen G, Yao Z. A blind watermarking scheme based on quantization for 3D models. Journal of Electronics & Information Technology, 2009, 31, 2963–2968.

[53] Zhao Q, Yin B. Sensitivity attack to watermarking schemes with nonparametric detection boundaries. Journal of Nanjing University of Science and Technology (Natural science), 2008, 32, 291–294.

[54] Van Schyndel R G, Tirkel A Z, Osborne C F. A digital watermark. IEEE International Conference on Image Processing, November 13–16, 1994, Austin, TX, 86–90.

[55] Abolghasemi M, Aghainia H, Faez K, Mehrabi M A. Steganalysis of LSB matching based on co-occurrence matrix and removing most significant bit plane. International Conference on Intelligent Information Hiding and Multimedia Signal Processing, August 15–17, 2008, Harbin, China, 1527–1530.

[56] Mielikainen J. LSB matching revisited. IEEE Signal Processing Letters, 2006, 13, 285–287.

[57] Ferreira R, Ribeiro B, Silva C, Liu Q, Sung A H. Building resilient classifiers for LSB matching steganography. IEEE International Joint Conference on Neural Networks, June 1–8, 2008, Hong Kong, 1562–1567.

[58] Pei S C, Cheng C M. Pallete-based color image watermarking using neural network training and repeated LSB insertion. The 13th IPPR Conference On Computer Vision, Graphic and Image Processing, August, 2000, Taiwan, 1–8.

[59] Ming C, Fan-fan L, Ru Z, Xin-xin N. Steganalysis of LSB matching in gray images based on regional correlation analysis. World Congress on Computer Science and Information Engineering, March 31–April 2, 2009, Los Angeles, CA, 490–494.

[60] Fu Y G, Shen R M. Color image watermarking scheme based on linear discriminant analysis. Computer Standards & Interfaces, 2007, 30, 115–120.

[61] Fu Y, Shen R, Lu H. Watermarking scheme based on support vector machine for color images. IET Electronics Letters, 2004, 40(16), 986–987.

[62] Kutter M, Jordan F, Bossen F. Digital signature of color images using amplitude modulation. Journal of Electronic Imaging, 1998, 7, 326–332.

[63] Yu P T, Tsai H H, Lin J S. Digital watermarking based on neural networks for color images. Signal Processing, 2001, 81, 663–671.

[64] Tsai H H, Sun D W. Color image watermark extraction based on support vector machines. Information Sciences, 2007, 177, 550–569.
[65] Huang P S, Chiang C S, Chang C P, Tu T M. Robust spatial watermarking technique for colour images via direct saturation adjustment. IEEE Proceedings on Vision, Image and Signal Processing, 2005, 152, 561–574.
[66] Kimpan S, Lasakul A, Chitwong S. Variable block size based adaptive watermarking in spatial domain. IEEE International Symposium on Communications and Information Technology, October 26–29, 2004, Thailand, 374–377.
[67] Verma B, Jain S, Agarwal D P, Phadikar A. A new color image watermarking scheme. Journal of Computer Science, 2006, 5, 37–42.
[68] Jun L, LiZhi L. An improved watermarking detect algorithm for color image in spatial domain. International Seminar on Future BioMedical Information Engineering, December 18, 2008, Wuhan, China, 95–99.
[69] Deng Y. Research on digital watermarking security. Beijing: Beijing University of Posts and Telecommunications, 2006.
[70] Podilchuk C I, Zeng W. Image-adaptive watermarking using visual models. IEEE Journal on Selected Areas in Communications, 1998, 16, 525–539.
[71] Cox I J, Kilian J, Leighton F T, Shamoon T. Secure spread spectrum watermarking for multimedia. IEEE Transactions on Image Processing, 1997, 6, 1673–1687.
[72] Niu X, Lu Z, Sun S. The embedding technique with color digital watermark. Acta Electronica Sinica, 2000, 28, 10–12.
[73] Wang X, Yang H. Adaptive 2-D color image watermarking based on DCT. Journal of Computer-Aided Design & Computer Graphics, 2004, 16, 243–247.
[74] Piva A, Bartolini F, Cappellini V, Barni M. Exploiting the cross-correlation of RGB-channels for robust watermarking of color images. IEEE International Conference on Image Processing, October 24–28, 1999, Kobe, 306–310.
[75] Hsieh M S, Tseng D C. Wavelet-based color image watermarking using adaptive entropy casting. IEEE International Conference on Multimedia and Expo, July 9–12, 2006, Toronto, ON, 1593–1596.
[76] Wang X, Yang H. Color digital watermarking based on integer lifting wavelet transform and visual masking. Journal of Computer-Aided Design & Computer Graphics, 2004, 16, 1240–1243.
[77] Jiang M, Chi X. Digital watermarking algorithm for color image based on IWT and HVS. Journal of Jilin University (Information Science Edition), 2007, 25, 98–102.
[78] Al-Otum H M, Samara N A. A robust blind color image watermarking based on wavelet-tree bit host difference selection. Signal Processing, 2010, 90, 2498–2512.
[79] Chen W Y. Color image steganography scheme using set partitioning in hierarchical trees coding, digital Fourier transform and adaptive phase modulation. Applied Mathematics and Computation, 2007, 185, 432–448.
[80] Tsui T K, Zhang X P, Androutsos D. Color image watermarking using the spatio-chromatic Fourier transform. 2006 IEEE International Conference On Acoustics, Speech and Signal Processing, May 14–19, 2006, Toulouse, 1553–1556.
[81] Yu Y U, Chang C C, Lin I C. A new steganographic method for color and grayscale image hiding. Computer Vision and Image Understanding, 2007, 107, 183–194.
[82] Tsai P, Hu Y C, Chang C C. A color image watermarking scheme based on color quantization. Signal Processing, 2004, 84, 95–106.
[83] Chou C H, Wu T L. Embedding color watermarks in color images. EURASIP Journal on Applied Signal Processing, 2003, 1, 32–40.

[84] Pei S C, Chen J H. Color image watermarking by Fibonacci lattice index modulation. Proceedings of the 3rd European Conference on Color in Graphics, Imaging, and Vision. Society for Imaging Science and Technology, 2006, 211–215.

[85] Tzeng C H, Yang Z F, Tsai W H. Adaptive data hiding in palette images by color ordering and mapping with security protection. IEEE Transactions on Communications, 2004, 52, 791–800.

[86] Lin C Y, Chen C H. An invisible hybrid color image system using spread vector quantization neural networks with penalized FCM. Pattern Recognition, 2007, 40, 1685–1694.

[87] Orchard M T, Bouman C. Color quantization of images. IEEE Transactions on Signal Processing, 1991, 39, 2677–2690.

[88] Jie N, Zhiqiang W. A new public watermarking algorithm for RGB color image based on Quantization Index Modulation. 2009 International Conference on Information and Automation, June 22–24, 2009, Zhuhai, Macau, 837–841.

[89] Chareyron G, Coltuc D, Trémeau A. Watermarking and authentication of color images based on segmentation of the xyY color space. Journal of Imaging Science and Technology, 2006, 50 (5), 411–423.

[90] Li J, Du W. Robustness watermarking technique for 2-D and 3-D medical image. Beijing: Intellectual Property Publishing House, 2011.

[91] Wang L, Zhang H, Ye D, Hu D. Information hiding technique and application. Hubei: Wuhan University Press, 2012.

[92] Liu J, Pang J. The matrix theory and method guidance. Hunan: Xiangtan University Press, 2008.

[93] Koschan A, Abidi M. Digital color image processing. Translated by Zhang Y. Beijing: Tsinghua University Press, 2010.

[94] Wang L, Guo C, Ye D, Li P. Information hiding technology experiment tutorial. Hubei: Wuhan University Press, 2012.

[95] Haralick R M, Shapiro L G. Glossary of computer vision terms. Pattern Recognition, 1993, 24 (1), 69–93.

[96] Robinson G S. Color edge detection. Optical Engineering, 1977, 16, 165479–165479.

[97] Gonzalez R C. Digital image processing. India: Pearson Education, 2009.

[98] Gilchrist A. Lightness, Brightness, and Transparency. New Jersey: Psychology Press, 2013.

[99] Zeki S. A Vision of the brain. Oxford: Blackwell Scientific, 1993.

[100] Kuehni R G. Color: An Introduction to practice and principles. New York: Wiley, 1997.

[101] Davidoff J. Cognition through color. Cambridge, MA: MIT Press, 1991.

[102] Poynton C A. A technical introduction to digital video. New York: Wiley, 1996.

[103] Pitas I, Tsalides P. Multivariate ordering in color image filtering. IEEE Transactions on Circuits and Systems for Video Technology, 1991, 1, 247–259.

[104] Plataniotis K N, Androutsos D, Vinayagamoorthy S, Venetsanopoulos A N. Color image processing using adaptive multichannel filters. IEEE Transactions on Image Processing, 1997, 6, 933–949.

[105] Zheng J, Valavanis K P, Gauch J M. Noise removal from color images. Journal of Intelligent and Robotic Systems, 1993, 7, 257–285.

[106] Adelson E H. The new cognitive neurosciences. Cambridge, MA: MIT Press, 2004.

[107] Hardeberg J. Acquisition and reproduction of color images: Colorimetric and multispectral approaches. Parkland, FL: Universal–Publishers, 2001.

[108] Giorgianni E J, Madden T E. Digital color management: Encoding solutions. New Jersey: Addison–Wesley Longman Publishing Co., Inc., 1998.

[109] Frey H. Digitale Bildverarbelin in Farbraumen. University Ulm, Germany, 1998.

[110] Foley J D, Van Dam A, Feiner S K, Hughes J F, Phillips R L. Introduction to computer graphics. Reading: Addison-Wesley, 1994.

[111] Lou D C, Tso H K, Liu J L. A copyright protection scheme for digital images using visual cryptography technique. Computer Standards & Interfaces, 2007, 29, 125–131.
[112] Fleet D J, Heeger D J. Embedding invisible information in color images. IEEE Transactions on Image Processing, October 26–29, 1977, Santa Barbara, CA, 532–535.
[113] Chan C K, Cheng L M. Hiding data in images by simple LSB substitution. Pattern Recognition, 2004, 37, 469–474.
[114] Qi X, Qi J. A robust content-based digital image watermarking scheme. Signal Processing, 2007, 87, 1264–1280.
[115] Usman I, Khan A. BCH coding and intelligent watermark embedding: Employing both frequency and strength selection. Applied Soft Computing, 2010, 10, 332–343.
[116] Lin S D, Shie S C, Guo J Y. Improving the robustness of DCT-based image watermarking against JPEG compression. Computer Standards & Interfaces, 2010, 32, 54–60.
[117] Baba S E I, Krikor L Z, Arif T, Shaaban Z. Watermarking of digital images in frequency domain. International Journal of Automation and Computing, 2010, 7, 17–22.
[118] Liu L S, Li R H, Gao Q. A new watermarking method based on DWT green component of color image. Proceedings of 2004 International Conference on Machine Learning and Cybernetics, August 26–29, 2004, Shanghai, China, 3949–3954,
[119] Liu K C. Wavelet-based watermarking for color images through visual masking. AEU-International Journal of Electronics and Communications, 2010, 64, 112–124.
[120] Shih F Y, Wu S Y. Combinational image watermarking in the spatial and frequency domains. Pattern Recognition, 2003, 36, 969–975.
[121] Thorat C G, Jadhav B D. A blind digital watermark technique for color image based on Integer Wavelet Transform and SIFT. Procedia Computer Science, 2010, 2, 236–241.
[122] Bohra A, Farooq O. Blind self-authentication of images for robust watermarking using integer wavelet transform. AEU-International Journal Electronics and Communications, 2009, 63, 703–707.
[123] Yuan Y, Huang D, Liu D. An integer wavelet based multiple logo-watermarking scheme. Proceedings of the 1st International Multi-Symposiums on Computer and Computational Sciences (IMSCCS'06), June 20–24, 2006, Hangzhou, Zhejiang, 175–179.
[124] Wang Y, Zhang H. A color image blind watermarking algorithm based on chaotic scrambling and integer wavelet. 2011 International Conference on Network Computing and Information Security, May 14–15, 2011, Guilin, China, 413–416.
[125] Acharya T, Chakrabarti C. A survey on lifting-based Discrete Wavelet Transform architectures. Journal of VLSI Signal Processing, 2006, 42, 321–339.
[126] Santiago-Avila C, Gonzalez M, Nakano-Miyatake M, Perez-Meana H. Multipurpose color image watermarking algorithm based on IWT and halftoning. ACS'10 Proceedings of the 10th WSEAS International Conference on applied computer science, 2010, 170–175.
[127] Sweldend W. The lifting scheme: A custom-design construction of bi-orthogonal wavelets. Journal of Applied & Computational Harmonic Analysis, 1996, 3, 186–200.
[128] Rivest R. The MD5 message digest algorithm. Internet RFC 1321, April 1992.
[129] Li L, Yuan X, Lu Z, Pan J S. Rotation invariant watermark embedding based on scale-adapted characteristic regions. Information Sciences, 2010, 180, 2875–2888.
[130] Golea N E H, Seghir R, Benzid R. A bind RGB color image watermarking based on Singular Value Decomposition. 2010 IEEE/ACS International Conference on Computer Systems and Applications (AICCSA), May 16–19, 2010, Hammamet, 1–5.
[131] Liu R, Tan T. SVD-based watermarking scheme for protecting rightful ownership. IEEE Transactions on Multimedia, 2002, 4, 121–128.
[132] Suresh G, Lalitha N V, Rao C S, et al. An efficient and simple Audio Watermarking using DCT-SVD. 2012 International Conference on Devices, Circuits and Systems (ICDCS), pp. 177–181, March 15–16 2012, Coimbatore, 177–181.

[133] Hiena T, Nakaoa Z, Chen Y W. Robust multi-logo watermarking by RDWT and ICA. Signal Processing, 2006, 86 (10), 2981–2993.
[134] Run R S, Horng S J, Lai J L, Kao T W, Chen R J. An improved SVD-based watermarking technique for copyright protection. Expert Systems with Applications, 2012, 39, 673–689.
[135] Bhatnagar G, Raman B. A new robust reference watermarking scheme based on DWT-SVD. Computer Standards & Interfaces, 2009, 31, 1002–1013.
[136] Bhatnagar G, Raman B. A new robust reference logo watermarking scheme. Multimedia Tools and Applications, 2011, 52, 621–640.
[137] Shih Y T, Chien C S, Chuang C Y. An adaptive parameterized block-based singular value decomposition for image de-noising and compression. Applied Mathematics and Computation, 2012, 218, 10370–10385.
[138] Mukherjee S, Pal A K. A DCT-SVD based robust watermarking scheme for grayscale image Proceedings of the International Conference on Advances in Computing, Communications and Informatic, 2012, New York, 573–578.
[139] Jia Y, Xu P, Pei X. An investigation of image compression using block Singular Value Decomposition. Communications and Information Processing Communications in Computer and Information Science, 2012, 288, 723–731.
[140] Basso A, Bergadano F, Cavagnino D, Pomponiu V, Vernone A. A novel block-based watermarking scheme using the SVD transform. Algorithms, 2009, 2, 46–75.
[141] Chandra D V S. Digital image watermarking using singular value decomposition. Proceedings of the 45th IEEE Midwest Symposium on Circuits and Systems, August 4–7, 2002, USA, 264–267.
[142] Huang F, Guan Z H. A hybrid SVD-DCT watermarking method based on LPSNR. Pattern Recognition Letters, 2004, 25, 1769–1775.
[143] Ghazy R A, El-Fishawy N A, Hadhoud M M, Dessouky M I. An efficient block-by block SVD-based image watermarking scheme. Proceedings of the 24th National Radio Science Conference, March 13–15, 2007, Cairo, 1–9.
[144] Ouhsain M, Hamza A B. Image watermarking scheme using nonnegative matrix factorization and wavelet transform. Expert Systems with Applications, 2009, 36, 2123–2129.
[145] Rao V S V, Shekhawat R S, Srivastava V K. A reliable digital image watermarking scheme based on SVD and particle swarm optimization. 2012 Students Conference on Engineering and Systems, March 16–18, 2012, Allahabad, Uttar Pradesh, 1–6.
[146] Dogan S, Tuncer T, Avci E, Gulten A. A robust color image watermarking with Singular Value Decomposition method. Advances in Engineering Software, 2011, 42, 336–346.
[147] Lei B Y, Soon I Y, Li Z. Blind and robust audio watermarking scheme based on SVD-DCT. Signal Processing, 2011, 91 (8), 1973–1984.
[148] Mohammad A A, Alhaj A, Shaltaf S. An improved SVD-based watermarking scheme for protecting rightful ownership. Signal Processing, 2008, 88, 2158–2180.
[149] Abdallah E E, Hamza A B, Bhattacharya P. Improved image watermarking scheme using fast Hadamard and discrete wavelet transforms. Journal of Electronic Imaging, 2007, 16, 0330201–0330209.
[150] Ganic E, Eskicioglu A M. Robust DWT-SVD domain image watermarking: Embedding data in all frequencies. Proceedings of the 2004 Workshop on Multimedia and Security. ACM, 2004, 166–174.
[151] Xu G, Wang L. Color image watermark algorithm based on SVD and Lifting Wavelet Transformation. Application Research of Computers, 2011, 28, 1981–1982.
[152] Yin C, Li L, Lv A Q, Qu L. Color image watermarking algorithm based on DWT-SVD. 2007 IEEE International Conference on Automation and Logistics, August 18–21, 2007, Jinan, China, 2607–2611.

[153] Golub G H, Van Loan C F. Matrix computations. Baltimore: Johns Hopkins University Press, 1989.
[154] Chang C C, Tsai P, Lin C C. SVD-based digital image watermarking scheme. Pattern Recognition Letters, 2005, 26, 1577–1586.
[155] Fan M Q, Wang H X, Li S K. Restudy on SVD-based watermarking scheme. Applied Mathematics and Computation, 2008, 203, 926–930.
[156] Li J. Research on digital image watermarking technology against geometric attacks. Nanjing: Nanjing University of Science & Technology, 2009.
[157] Chung K L, Yang W N, Huang Y H, Wu S T, Hsu Y C. On SVD-based watermarking algorithm. Applied Mathematics and Computation, 2007, 188, 54–57.
[158] Chen W, Quan C, Tay C J. Optical color image encryption based on Arnold transform and interference method. Optics Communications, 2009, 282, 3680–3685.
[159] Yin Z H, Zhou Y D, Gao D H, Zhang J Y, Lin X, Han Y N. The anti-zoom capability of digital watermarking based on difference feature point grid. Journal of Air Force Engineering University (Natural Science Edition), 2009, 10, 76–80.
[160] Schur I. On the characteristic roots of a linear substitution with an application to the theory of integral equations. Mathematische Annalen, 1909, 66, 488–510.
[161] Li X, Fan H. QR factorization based blind channel identification and equalization with second-order statistics. IEEE Transaction on Signal Processing, 2000, 48, 60–69.
[162] Moor B, Dooren P. Generalizations of the singular value and QR decompositions. SIAM (Society for Industrial and Applied Mathematics) on Matrix Analysis and Applications, 1992, 13, 993–1014.
[163] Wang P, Liu M, Gong D. The digital image watermarking method for authentication based on QR decomposition. China, 2011100799133, March 03, 2011.
[164] Yashar N, Saied H K. Fast watermarking based on QR decomposition in Wavelet domain. 2010 6th International Conference on Intelligent Information Hiding and Multimedia Signal Processing, October 15–17, 2010, Darmstadt, 127–130.
[165] Song W, Hou J J, Li Z H, Huang L. Chaotic system and QR factorization based robust digital image watermarking algorithm. Journal of Central South University of Technology, 2011, 18, 116–124.
[166] Su Q, Niu Y, Wang G, Jia S, Yue J. Color image blind watermarking scheme based on QR decomposition. Signal Processing, 2014, 94, 219–235.
[167] Yang H Y, Wang X Y, Wang P, Niu P P. Geometrically resilient digital watermarking scheme based on radial harmonic Fourier moments magnitude. International Journal of Electronics and Communications, 2015, 69, 389–399.
[168] Niu P P, Wang X Y, Yang Y P, Lu M Y. A novel color image watermarking scheme in nonsampled contourlet-domain. Expert Systems with Applications, 2011, 38, 2081–2098.
[169] Shao Z, Duan Y, Coatrieux G, Wu J, Meng J, Shu H. Combining double random phase encoding for color image watermarking in quaternion gyrator domain. Optics Communications, 2015, 343, 56–65.
[170] Guo J, Prasetyo H. Security analyses of the watermarking scheme based on redundant discrete wavelet transform and singular value decomposition. International Journal of Electronics and Communications, 2014, 68, 816–834.
[171] Wang X, Wang C, Yang H, Niu P. A robust blind color image watermarking in quaternion Fourier transform domain. Journal of Systems and Software, 2013, 86, 255–277.
[172] Naderahmadian Y, Hosseini-Khayat S. Fast and robust watermarking in still images based on QR decomposition. Multimedia Tools and Applications, 2013, 72, 2597–2618.
[173] Bhatnagar G, Wu Q M. Biometrics inspired watermarking based on a fractional dual tree complex wavelet transform. Future Generation Computer Systems, 2013, 29, 182–195.

[174] Seddik H, Sayadi M, Fnaiech F, Cheriet M. Image watermarking based on the Hessenberg transform. International Journal of Image and Graphics, 2009, 9, 411–433.

[175] Su Q, Niu Y, Zou H, Zhao Y, Yao T. A blind double color image watermarking algorithm based on QR decomposition. Multimedia Tools and Applications, 2014, 72 (1), 987–1009.

[176] Golub G H, Loan C F V. Matrix computations. Baltimore: Johns Hopkins University Press, 1996.

Index

Bei Fragen zur Produktsicherheit wenden Sie sich bitte an:
If you have any questions regarding product safety,
please contact:

Walter de Gruyter GmbH
Genthiner Straße 13
10785 Berlin
productsafety@degruyterbrill.com